Basic
Laser Raman
Spectroscopy

Basic
Laser Raman
Spectroscopy

by Jack Loader
University of Southampton

PUBLISHED BY	**HEYDEN & SON LTD**
IN CO-OPERATION WITH	**SADTLER RESEARCH LABORATORIES INC.**

Heyden & Son Ltd., Spectrum House, Alderton Crescent, London, NW4

Heyden & Son Inc., 1220 Lexington Avenue, New York, N.Y. 10028, USA

Heyden & Son GmbH, Steinfurter Str. 45, 4440 Rheine/Westf., Germany

Co-published for exclusive distribution
in the USA by
Sadtler Research Laboratories Inc.,
3314–20 Spring Garden Street,
Philadelphia Pa. 19104

Library of Congress Catalog Card No. 75–111032
SBN 85501 020 7

Made and printed in Northern Ireland
at the Universities Press, Belfast.

Contents

 # Preface

This small book is intended as an aid to the growing numbers of chemists who are coming into or returning to Raman spectroscopy as the result of the introduction of the laser as a monochromatic light source and the new range of commercial spectrometers available as a result of this. The section on the Raman effect is an attempt to explain it in non-mathematical terms so that the instrument operator has some idea of both what is happening to the sample in the instrument while he is running the spectrum, and what the spectrum means when he has obtained it.

The chapter on the instruments is to acquaint the reader with the sampling techniques that have been developed by the users for chemical problems and is intended as an extension to the relevant handbooks. It will, of course, also be of use to the chemist who wishes to submit a sample to be run on an instrument of which he has no personal knowledge.

I wish to express my sincere thanks to Dr. P. J. Hendra who suggested the original idea and with whom I have had many helpful discussions; to Professor I. R. Beattie for the extended use of the Cary 81; to Dr. D. M. Adams for allowing me to see his Perkin-Elmer LR 1 and his sampling methods; and to Dr. H. J. Clase for showing me the Coderg PH1 and the cells he has developed and for supplying one of the spectra in the book. I am indebted to several of my colleagues who have read sections of the manuscript and have made many useful suggestions, to my wife, Anne, for typing the manuscript and to Mrs. J. M. Hendra for drawing some of the diagrams.

JACK LOADER
Southampton, 1969

1

The Raman effect

HISTORICAL

The effect was predicted theoretically by Smekal[1] in 1923 and was discovered by Sir C. V. Raman in 1928. For this discovery and his subsequent work on the effect, he was awarded the Nobel prize for Physics in 1930.

In the course of his studies of molecular light scattering Raman[2] directed a beam of sunlight, which had been condensed by a telescope, through a liquid. He observed that when two complementary light filters, blue-violet and yellow-green, were placed between the telescope and the container of liquid the track of the light was completely extinguished. If, however, the yellow filter was placed behind the container, between it and the observer's eye, the track again became visible indicating that the light had been modified in some way by its passage through the liquid.

In a paper[3] submitted to the Royal Society later in 1928 Raman showed photo-graphically recorded spectra of several liquids including benzene and carbon tetra-chloride. The spectrum of the latter was recorded using the 4358 Å lines of mercury from a 3000 c.p. lamp, a 600 ml vacuum distilled sample and a 25 hour exposure of the photographic plate.

Since that time there has been a gradual improvement in instrumentation and sources. A great many sources have been tried, but have all suffered from instability and trouble with unwanted emission lines of comparable intensity to the desired exciting line. The most important development in emission lamps came in 1952 when Welsh et al.[4] introduced the Toronto arc mercury lamp which consists of a four-turn helix of pyrex tubing. It can radiate as much as 50 watts of energy in the 4358 Å emission lines of mercury, but only a very small fraction of this excites the Raman spectrum.

In 1962 Porto and Wood[5], and Stoicheff[6] employed pulsed ruby masers to obtain the Raman spectra of some strongly scattering liquids. These experiments were slow and difficult, but during the last few years tremendous advances have been made in lasers and in spectrometer design, with the result that Raman spectroscopy has been completely revolutionised.

By the use of lasers as high power monochromatic sources, the complete Raman spectrum of 10 μl of a liquid sample straight from the bottle can be recorded in less than twenty minutes. Good quality spectra have been recorded from solids, gases and single crystals of all colours and from as little as 0·1 μl of liquid without great difficulty.

Since the advent of the laser, exposure times of the order of those used by Raman have been reserved for the study of high resolution gas phase spectra. Recently, however, Barrett and Adams[7] have used a pulse counting technique to record, photoelectrically, very weak Raman spectra. They obtained, in under three hours, an excellent vibration-rotation spectrum of a 10^{-7} ml sample of nitrogen gas at atmospheric pressure.

GENERAL CONSIDERATIONS

When light passes through a transparent medium a small fraction of it is scattered by the molecules. This scattering is composed of two parts:

(a) Rayleigh scattering – here the light is elastically scattered in all directions by the molecules.

(b) Raman scattering – this is by far the weaker of the two effects. The light is inelastically scattered by the molecules, and it carries away with it information about the vibrational and rotational energy levels of the molecules.

If the light illuminating the sample is monochromatic and the scattered light is examined with a spectrometer, a series of emission lines will be seen. The strongest line appears at the frequency of the exciting monochromatic light and is due to Rayleigh scattering. Symmetrically placed on either side of the Rayleigh line are a number of very much weaker lines, these being the Raman emissions. The Raman emissions on the low frequency side of the Rayleigh line are called the Stokes lines, and are of higher intensity than the anti-Stokes lines, which lie on the high frequency side.

This intensity difference between the two sets of lines arises because of the population difference between the different vibrational energy levels of the molecule. At room temperature there will be many more molecules in the ground vibrational state than in the higher vibrational states, and therefore the incoming light is more likely to interact with a molecule in the ground state and excite it to a higher vibrational state than it is to collide with a molecule in one of the higher energy states, causing it to lose energy and fall back to the ground state. If the light excites the molecule to a higher vibrational state it will lose energy and appear with a lower frequency, conversely, if it brings about a downward transition it will gain energy and appear at a higher frequency.

The displacements of these lines from the Rayleigh line ($\Delta\nu$), measured in wavenumbers (cm^{-1}) are found to correspond to the frequencies of the molecular vibrations. These frequencies can sometimes be measured by absorption spectroscopy in the infrared.

The infrared and Raman spectra are not alternatives, both being necessary if the maximum amount of information is to be obtained about a molecule.

The activity of a particular vibrational mode in the infrared region is dependent upon whether or not there is a change in the dipole moment during the vibration. For a mode to be Raman active there must be a change in the polarisability of the molecule during the vibration. This change can be considered as being a change in the shape of the electron cloud surrounding it.

If we consider the carbon disulphide molecule CS_2, this is a linear molecule with a centre of symmetry and no permanent dipole moment. We know from elementary vibration theory that a linear molecule with N atoms has $3N - 5$ vibrational modes, thus for carbon disulphide we would expect $3 \times 3 - 5 = 4$ modes.

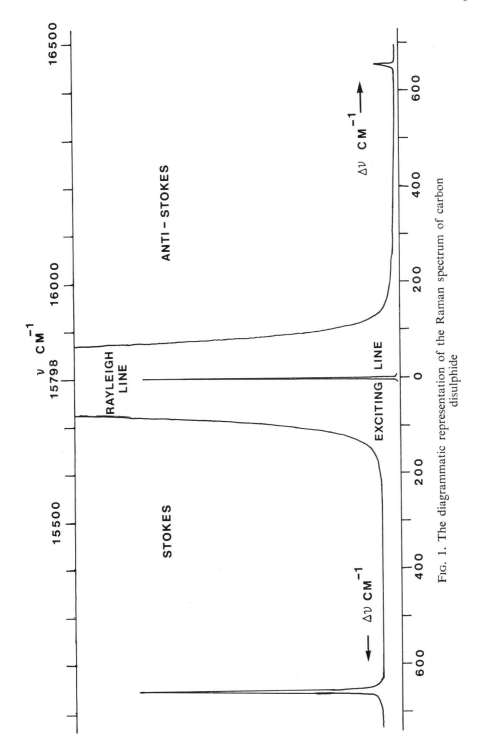

Fig. 1. The diagrammatic representation of the Raman spectrum of carbon disulphide

These modes can be illustrated pictorially as follows:

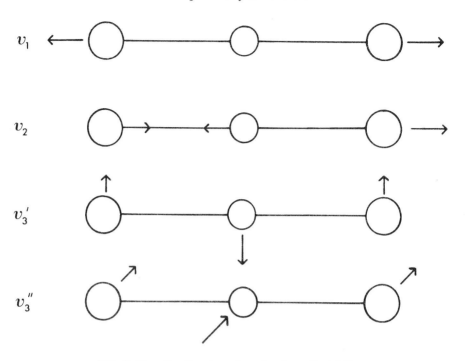

FIG. 2. The vibrational modes of carbon disulphide

The ν_3' vibration occurs in the plane of the paper and the ν_3'' vibration in the plane at right angles to the paper. In space these two are indistinguishable and they are always considered together as the one 'doubly degenerate' vibration.

We know that the sulphur atoms will have a slightly higher electron density, i.e. a slightly higher negative charge, than the carbon atoms. In ν_2 and ν_3, therefore, there will be a change in the dipole moment, since during these vibrations the centres of highest positive and negative charge will move in such a way that the electrical centre of the molecule is displaced from the carbon atom. These vibrations will be observed in the infrared spectrum of the molecule.

In the symmetric stretching mode ν_1 there will be no change in the dipole moment, as the two negative centres will move equal distances in opposite directions from the positive centre. The electron cloud around the molecule will expand, however, and thus there will be a change in the polarisability, and the vibration will be Raman active.

It is more difficult to see intuitively that there will not be a change in the polarisability during the vibrations ν_2 and ν_3. This is found, however, to be the case.

It is a general rule for a molecule with a centre of symmetry that a particular vibration cannot appear in both the infrared and Raman spectra.

For any molecule, the number of modes that are active in the Raman or infrared can be predicted by the methods of group theory. For details of this and a more comprehensive explanation of the Raman effect, the reader is referred to the standard texts on the subject (see Bibliography).

POLARISATION MEASUREMENTS

If the light illuminating the sample is polarised as well as being monochromatic and an analyser is placed between it and the spectrometer, the Raman lines will be observed to have different intensities for different directions of polarisation of the incident beam. If $I_{\parallel}$ is the intensity of a Raman line with the incident light polarised parallel to the direction in which the analyser passes the maximum amount of light, and $I_{\perp}$ is the intensity of the line with the incident radiation polarised perpendicular to that direction, then it is possible to define the depolarisation ratio, ρ, as follows:

$$\rho = \frac{I_{\perp}}{I_{\parallel}}$$

It has been shown theoretically for a sample illuminated with a laser that the maximum value of ρ is 0·75. A line having this value arises only from non-totally symmetric vibrations and is said to be depolarised.

A line with a depolarisation ratio of less than 0·75 is said to be polarised and it can arise only from a totally symmetric vibration. By measuring the depolarisation ratios of all the lines in the Raman spectrum of a liquid sample, it is possible to unambiguously assign lines due to totally symmetric modes of vibration. This does not, however, apply to polycrystalline samples because multiple internal reflection by the crystals depolarises the light to some unknown and unpredictable extent.

2
Sample handling

Each of the commercial instruments have their own sampling procedures and these are fully described in the relative handbooks. In general, however, laboratories using the instruments have devised additional methods for use with their particular models, and these will be discussed separately after some general considerations.

GENERAL CONSIDERATIONS

1. Laser radiation

It is extremely dangerous to look directly down a laser beam, as serious eye damage can result from such action. Accidental viewing of the laser beam is unlikely in most spectrometers as the instrument manufacturers, aware of the dangers, have taken measures to prevent this happening.

However, sample alignment in the spectrometer done with an unattenuated laser beam is much more hazardous, as eye damage can result from accidental reflection of the laser beam from a polished surface.

All operations of this type where such risks are present are best carried out in the highest possible level of room light so that the pupil of the eye is small, thereby minimising any damage done.

Safety goggles are available and they give protection against reflected light but are not to be recommended for looking straight down the beam. Some of the early ones were so deeply coloured that it was impossible to work without a high power white light source illuminating the sample, but now goggles are available based on broad-band interference filters and these are much more practical.

'Spectroguard' goggles, marketed by Spectrolabs, 12484, Gladstone Avenue, Sylmar., Calif. 91342, USA, are of about the same colour as normal sun glasses yet they provide simultaneous protection against all laser lines used in Raman spectroscopy except the yellow (5682 Å) line of the krypton ion laser.

2. Cells

All cells should be clean and free from grease and fingerprints as these can cause a considerable increase in the fluorescent background of the spectrum and from air bubbles which increase the scatter of the laser beam, thereby reducing the excitation efficiency and increasing the amount of laser radiation reaching the monochromator and the operator's eyes.

The outside of the cell should be wiped with a tissue moistened with chloroform or acetone to remove as much dirt as possible before use.

3. Fluorescence

Sample fluorescence is much less of a problem with the laser as a source than it was with the mercury arc.

It has been shown that the fluorescent background produced by some samples will decay gradually with time if they are left in the laser beam. It is therefore good practice to wait some time before recording the spectrum of a strongly fluorescent material as the background may reduce considerably.

4. Care of photomultiplier tubes

The outside of the photomultiplier must be clinically clean and free from all traces of grease and fingerprints. Failure to keep the tube clean may result in its being excessively noisy.

Under no circumstances should an e.h.t. voltage be applied to the tube when it is in room light as this will almost certainly destroy it. Care should also be exercised when scanning near the exciting line to ensure that the tube is not exposed to an excessively high intensity of light. If the latter type of exposure does occur and the tube is not destroyed it will certainly need to be kept at full e.h.t. potential in the dark for several hours before the dark current returns to a reasonable value.

If possible the spectrometer should be switched on some time before it is wanted so that the phototube will have time to settle down to its normal operating conditions; better still, the e.h.t. voltage should be left on all the time.

Under no circumstances should the instrument be switched off for a short time between users as the phototube will take up to an hour to recover fully.

CARY 81

This instrument is manufactured by Cary Instruments Inc., 2724 South Peck Road, Monrovia, California 91016, USA, and supplied in the United Kingdom by Cary Instruments, Russell House, Molesey Road, Walton-on-Thames, Surrey.

It is normally supplied with a Spectra Physics 125A 50 mW helium neon laser but it can take almost any commercially available c.w. laser. 180° and 90° illumination is used together with an image slicer which gives the spectrometer a very high light gathering power. A double monochromator with a resolution of 0.5 cm^{-1} at 6328 Å is used to cover the region from 3985 Å to 8470 Å (25100 to 11800 cm^{-1}).

Detection is by S-20 response photomultipliers in conjunction with a phase sensitive detector and a.c. amplification. The spectrum is scanned linearly with respect to wavenumber and recorded on a coupled strip chart recorder which may be calibrated automatically by a marker every 100 cm^{-1}. The pen response control is continuously variable, allowing it to be altered during a scan without the risk of introducing switching transients. A zero suppression control is incorporated to suppress unwanted background signals.

The $\Delta\nu$ cm^{-1} counter has an external adjustment so that it may be readily set for any exciting line.

It is supplied with holders for solids, and with cells and holders for liquids. A cell for obtaining spectra from gases is also available.

Liquids

Several methods have been developed depending on the data required and on the nature of the liquid.

The greatest sensitivity and best signal-to-noise ratio is obtained using a Cary capillary cell. Only slightly inferior results, however, are obtained from a Pyrex glass capillary tube with a rounded end, provided that good optical contact with the hemispherical lens is maintained by means of a spot of glycerol. These tubes are of the type normally used for measuring the melting points of organic compounds, and should be thin-walled with an internal diameter of about 1 mm. The spectral intensity obtained increases with length up to about 15 cm; at lengths in excess of this the increase is much less rapid and is often outweighed by the difficulty in filling.

Liquids can also be examined in any thin-walled glass vessel which is small enough to be brought into contact with the hemispherical lens, e.g. small polythene-capped sample tubes. For use with compounds which, for a variety of reasons, must not be exposed to the air, there is the thin-walled sealed ampoule; this can have any diameter in excess of about 0·5 cm. This illustrates the great advantage of Raman over infrared spectroscopy in that a compound can be made or purified on a vacuum line, distilled into an ampoule attached to the line, sealed off and its spectrum obtained in the ampoule without any trouble.

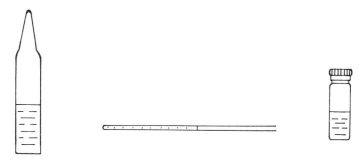

FIG. 3. Liquid cells for use with the Cary 81 spectrometer

In both of the last two methods the tube is viewed through the side and the sample is adjusted until the laser beam passing through the tube strikes the centre of the plug in the rear of the tower. While both of these methods are very convenient they do suffer from the disadvantage that the spectral intensity is about 35 times less than that obtained from a capillary tube.

Polarisation measurements for liquids

Although the best signal-to-noise ratio in the spectrum of a liquid is obtained when it is contained in a capillary cell, however, this method does not give good polarisation results. The cell relies for its high efficiency on the multiple reflection of the Raman light from the walls, bringing a large proportion of it to the collector lens. These multiple reflections result in the light becoming depolarised to some considerable extent and this means that the measured value of the depolarisation ratio may be considerably different from the theoretically predicted value.

If the spectrum has to be obtained with the liquid in a capillary tube, the tube should first be calibrated with carbon tetrachloride, since the true depolarisation ratios for this compound are well known. The line at $\Delta\nu = 318$ cm^{-1} has a depolarisation ratio of 0·75 and according to a recent paper by Murphy, Evans and Bender[8] the depolarisation ratio of the $\Delta\nu = 458$ cm^{-1} line is 0·0039.

This is illustrated by the calibration of the capillary tube used to obtain the carbon tetrachloride spectrum shown in Appendix C (see p. 71). The observed depolarisation ratios are $\rho_{459} = 0·535$ and $\rho_{318} = 0·943$.

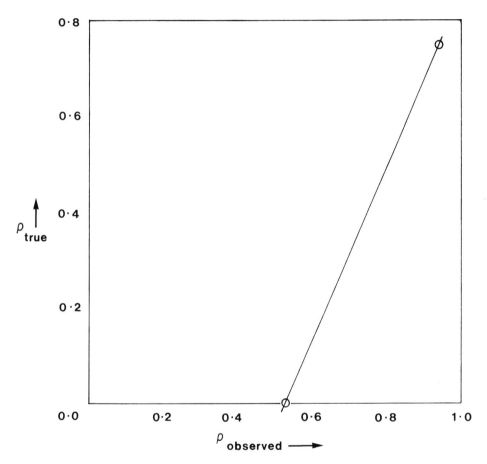

FIG. 4. The calibration curve of a capillary cell for use for polarisation measurements with the Cary 81 spectrometer

Allkins and Lippincott[9] have described a cell which gives very good polarisation results. The liquid is contained in a 1 cm path ultraviolet cell and is viewed through one side. The opposite side has a plane mirror against it to improve the spectral intensity which is down by a factor of about twenty from that of the capillary cell.

Coloured liquids

The calibration curve obtained in the way outlined above for a capillary tube will not necessarily apply to a coloured liquid. This is due to the absorption of the primary

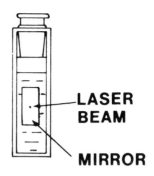

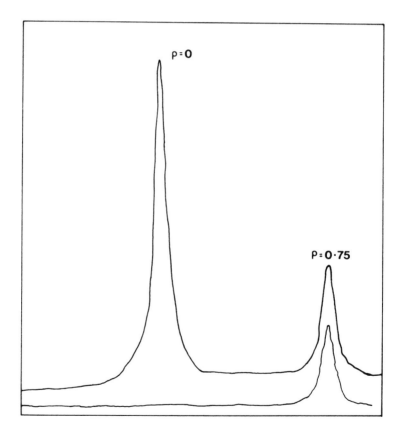

FIG. 5. A polarisation cell due to Allkins and Lippincott. Part of a spectrum of carbon tetrachloride obtained in the cell

beam and the Raman radiation by the liquid. In many cases where exact polarisation data are required these difficulties can often be by-passed by the inclusion of an internal standard whose depolarisation ratios are known. Murphy, Evans and Bender[8], however, have shown that the depolarisation ratio of the totally polarised line of carbon tetrachloride ($\Delta\nu = 459$ cm^{-1}) is dependent upon the molecular environment.

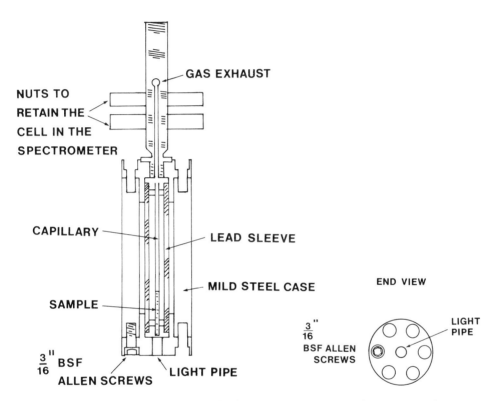

FIG. 6. An explosion-proof cell devised by Hendra and Mackenzie for use with the Cary 81 spectrometer

They have shown that the value of the depolarisation ratio of carbon tetrachloride also varies slightly from solvent to solvent in solution, and this must be borne in mind when making measurements using it as an internal standard. Similar considerations may well apply to other compounds that could be used.

Solids

The facile study of a wide range of solids is one of the outstanding features of this instrument. It has been found that spectra can be obtained from almost any solid provided that it can be brought close to the hemispherical lens, any gaps being bridged with glycerol. This method has been used for obtaining the spectra of polymers (see the polythene spectrum included in the spectra section), minerals and coatings.

Unreactive powders can be pressed into a pellet and placed directly in contact with the hemispherical lens. A spot of glycerol, however, must be applied to the pellet to establish good optical contact and this results in the loss of the sample.

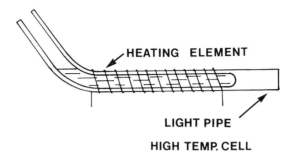

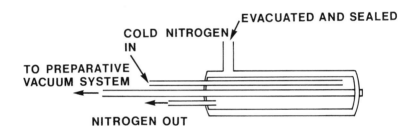

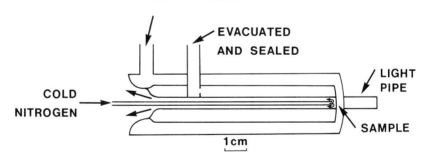

FIG. 7. High and low temperature cells for use with the Cary 81 spectrometer

(by courtesy of Dr. K. M. S. Livingston and Dr. G. A. Ozin)

Perhaps a better way when a quantity of the sample is available is to examine it in a polythene-capped sample tube viewed through the side.

Small quantities of solid can be run in a narrow bore tube with an optical flat on the end – this can of course be sealed onto a gas line or filled in a dry box. Provided that the optical flat is fairly thin there is very little difference in intensity between spectra run using such a cell and those run in a sample tube.

The tube with the optical flat has been used to obtain the spectra of species adsorbed onto catalyst surfaces.[10]

Explosive compounds

Hendra and Mackenzie[11] have developed an explosion-proof cell for use with this instrument. With its use they have obtained the Raman spectrum of pure liquid nitrogen trichloride. This is an extremely dangerous compound and during its preparation a small quantity exploded, completely destroying the whole preparative apparatus.

High and low temperature operation

In both of these the sample is held some way from the hemispherical lens and the light is conducted to and from it via a light pipe. This can either be one of the commercial types* or simply a short piece of Pyrex rod. In general the former give much better results. Commercial light pipes are limited to 450°C, and therefore above this temperature researchers have resorted to leaving an air gap between the heated end of the sample tube and the pre-slit optics of the spectrometer. There is a considerable loss of intensity, but a spectrum can be obtained in favourable cases.

The low temperature solid and liquid cells have vacuum jackets to minimise the heating of the sample and are maintained at the required temperature by a stream of cold nitrogen gas.

Another low temperature cell designed especially for the Cary 81 has been described by Carlson.[12] With the aid of this cell high quality spectra were obtained from solids and condensed vapours at temperatures down to $-135°C$.

CODERG CH 1 AND PH 1

These instruments are manufactured by Coderg, 15 Impasse Barbier, 92, Clichy, France. There are available a number of 'transfer plates' which fit both instruments and enable the operator to run spectra of solids and liquids and to work under conditions of variable temperature down to 4°K using 90° sample illumination.

The CH 1 is the so-called Chemical Version. It uses a double monochromator but the two halves are not mechanically coupled. The second half, which can contain either a grating or a prism, acts only as a filter to eliminate ghosts and stray light. It is adjusted manually. The spectrometer has a resolution of 1 cm^{-1} and allows work up to 50 cm^{-1} from the exciting line.

The PH 1 is the so-called Physics Version and it has a coupled double monochromator which cuts down the scattered light in the instrument and allows work much closer to the exciting line. It has a resolution of 0·5 cm^{-1} at 6328 Å which makes the resolution of the carbon tetrachloride band at $\Delta \nu = 459$ cm^{-1} into its components an easy operation.

In both instruments the Raman light is detected by a high performance photomultiplier tube and a d.c. amplifier. The spectrum is scanned linearly by wavenumber and displayed on a non-coupled strip recorder which is calibrated by a marker every 50 cm^{-1}. A light on the control panel indicates that the correct combination of scan speed, slit width and response time has been set by the operator.

* Available, for example, from Messrs Barr & Stroud Ltd., Caxton St., Anniesland, Glasgow W.3.

The instruments are normally supplied with a Spectra Physics Model 125 A 150 mW helium-neon laser or a Model 141 argon ion laser but can accept almost any commercially available c.w. laser operating between 4000 and 7000 Å.

Liquids

It has been found that excellent spectra can be obtained from a cell that is very easily made. A length of thin-walled capillary tube of internal diameter about 1 mm is heated until there is a lump of molten glass on the end. It is then pressed against a carbon block to produce a flattened end. In order for the cell to fit into the standard transfer plate, where it is retained by a piece of cork cut to size, it should not be longer than 22 mm.

FIG. 8. Liquid cell for use with the Coderg spectrometers

Polarisation measurements

As a result of the sampling method used in this instrument, the maximum value of the depolarisation ratio is 6/7. This difference arises because the measured quantity is not the usual $I_\perp/I_\parallel$ but $2I_\perp/I_\perp + I_\parallel$.

Solids

A solid sample cell that will give excellent results is shown in Fig. 9.

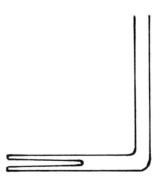

FIG. 9. Solid cell for use with the Coderg spectrometers

This cell is made as follows:

  (i) A length of thin-walled capillary tube of about 2 mm internal diameter is pulled out to a sharp point.

(ii) This is then sealed into another piece of the same tube and the first tube is then cut off to the left of the joint (as shown in Fig. 10).

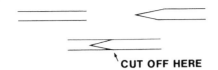

CUT OFF HERE

FIG. 10. Details of the construction of the Coderg solid cell

(iii) The cell is then bent so that it is less than 22 mm long and will fit into the standard transfer plate. It can be readily sealed to a gas line or filled in a dry box and can thus be used for reactive compounds.

PERKIN-ELMER LR 1

This instrument is produced by the Perkin-Elmer Corporation, 750 Main Avenue, Norwalk, Connecticut 06852, USA, and sold in the United Kingdom by Perkin-Elmer Ltd., Beaconsfield, Bucks. It is fitted with a Perkin-Elmer 6 mW helium-neon laser and uses 90° sample illumination.

A double-pass single monochromator with a resolution of $2 \cdot 5$ cm^{-1} at 6328 Å is used and the spectrum is scanned linearly by wavelength. The Raman light is detected with a S 20 response photomultiplier tube in conjunction with an a.c. amplifier and the spectrum is displayed on a non-coupled recorder which is calibrated by a marker.

This was the first commercial instrument to appear on the market, first being shown at the Ohio State University in 1964. Production of this instrument has now ceased as it has been replaced by the LR3 (see p. 52).

Liquids

Samples that have been produced or purified on a gas line can be run in a sealed ampoule which has been bent to fit into the sample area of the instrument (Fig. 14).

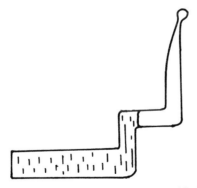

FIG. 11. Sealed liquid cell for use with the Perkin-Elmer LR 1

SPEX LASER RAMAN SYSTEMS

These instruments are manufactured by Spex Industries Inc., 3880 Park Avenue, Metuchen, N.J. 08840, USA. The British agents are Glen Creston, The Red House, 37 The Broadway, Stanmore, Middlesex.

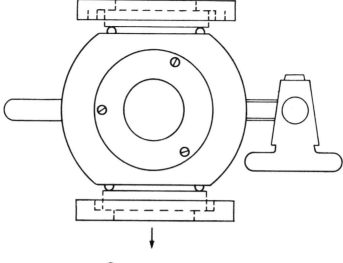

Spectrometer

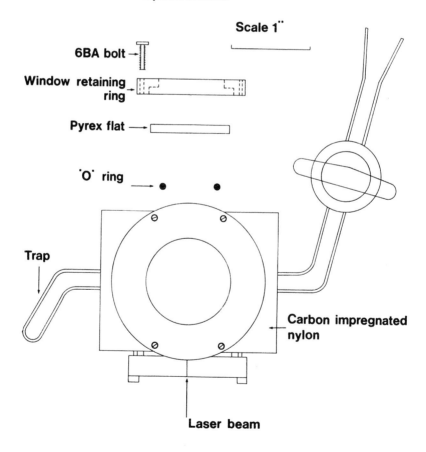

FIG. 12. Gas cell for use with Spex

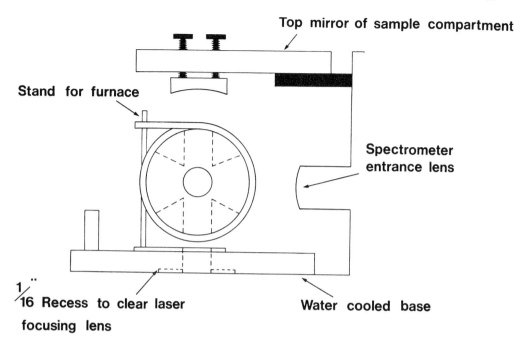

Top mirror of sample compartment

Stand for furnace

Spectrometer
entrance lens

$\frac{1}{16}$" Recess to clear laser
focusing lens

Water cooled base

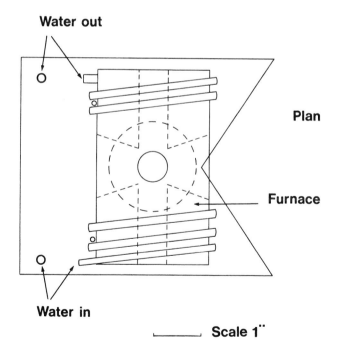

Water out

Plan

Furnace

Water in

Scale 1"

FIG. 13. Furnace for use with Spex

They are available either as a collection of modules – laser, power supply, double monochromator, detection electronics, and recorder, or else as a 'Ramalog' where all the modules are built into a specially constructed bench.

There is a choice of two double monochromators, the Model 1400 which scans linearly with respect to wavelength, or the Model 1401 which scans linearly with respect to wavenumber. Otherwise they are similar and have a resolution of $0\cdot2$ cm^{-1} at 6328 Å and cover the range 1800 to 15000 Å (55560 to 6667 cm^{-1}).

The instrument is normally supplied with a Spectra Physics 125A 50 mW helium-neon laser, but it can accept almost any commercially available c.w. laser. The laser beam enters the sample area from below and 90° illumination is used.

Detection is by a photomultiplier with an S-20 response in conjunction with either photon counting or d.c. amplification. Attachments are available to cool the photo-multiplier to either $-20°$C thermoelectrically, or to 100° K using a cryostat.

An event marker is available as an accessory. This plug-in unit will facilitate the use of a digital encoder with the instrument.

Sample holders for opaque solids, single crystals and liquids are supplied.

Gases

The cell illustrated in Fig. 12 has been designed and built by C. J. Vear. The basic material of construction is carbon impregnated nylon and the Pyrex windows are retained by dural rings. The nylon is sufficiently strong to allow holes for the ring retaining screws to be tapped directly into it. The glass trap and tap are held in position by Apiezon W40 wax.

The cell fits into the Spex sample area without disturbing any of the sample compartment optics.

Using this cell, high quality gas phase spectra have been obtained from many small molecules.

High temperature furnace

This furnace (Fig. 13) was designed by Dr. J. Horder and can be used at temperatures of up to 1000° C.

The sample area back mirror and its platform has to be removed but the upper mirror remains in position.

The furnace is constructed in a $1\frac{3}{4}''$ diameter brass tube and is wound with nichrome V. The ends and base are water cooled to prevent damage to the instrument.

The sample is contained in a 12 mm O.D. tube which is made either of Pyrex or silica depending on the operating temperature. By sealing the sample into the tube under a pressure of inert gas, Raman spectra have been obtained at moderate pressures ($<$ 30 atmospheres) and elevated temperatures.

3

Calibration of the spectrometer

STANDARD CALIBRATION

The spectrometer can be calibrated readily to high precision by using the emission spectrum of neon. Neon lamps are available from many suppliers and the emission lines cover the whole of the electromagnetic spectrum normally used in Raman spectroscopy (3100 to 8900 Å) with an average spacing of about 20 Å.

The lasers most commonly used are the helium-neon (6328 Å) and the argon ion (4880 Å mainly but also 5145 Å and other weaker lines). Tables 2 and 3 below show the neon lines which will be found on the Stokes (i.e. lower frequency) side of the laser lines. The shifts in wavenumbers from the exciting line are also given.

The measurement of the wavelengths of the neon emission lines was made by Burns, Adams and Longwell[13] interferometrically in standard air. The observed values were corrected to frequencies *in vacuo* but later work of Edlén[14] on the refractive index of air showed these corrections to be incorrect. In Tables 2 and 3 the frequencies listed have been recalculated using Edlén's formula from Burns, Adams & Longwell's measured wavelength values. The best available values for the laser wavelengths and frequencies are given in Table 1.

TABLE 1

Calibration source	Wavelength (Å)	Frequency (cm^{-1}_{vac})	Calibration source	Wavelength (Å)	Frequency (cm^{-1}_{vac})
Helium-neon	6328·1646	15,798·002	Krypton ion	4619·17	21,642·85
Argon ion	4879·865	20,486·684		4680·45	21,359·49
	5145·27	19,429·91		4762·44	20,991·77
	4579·36	21,830·99		4825·18	20,718·83
	4657·95	21,462·66		5208·32	19,194·70
	4726·89	21,149·64		5308·68	18,831·83
	4764·88	20,981·02		5681·92	17,594·80
	4965·09	20,135·00		6471·00	15,449·29
	5017·17	19,926·00		6764·57	14,778·83

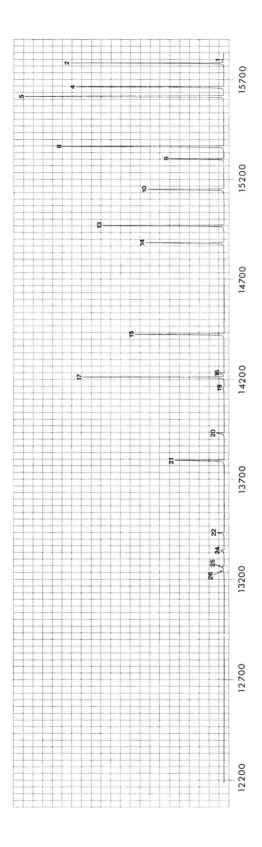

Light source: Neon lamp
Reference dynode: 3
Raman dynode: 4
Spectral slit width: 1 cm^{-1}
Scan speed: 250 cm^{-1}/min
Slit length: 10 cm
Sensitivity: 100
Single/Double slits: Single
Pen response: 0·1 sec

TABLE 2. Calibration of laser Raman spectrometers for
helium-neon excitation

	Wavelength (Å)	Frequency (cm_{vac}^{-1})	Apparent Raman shift (cm_{vac}^{-1})
1	6328·1646	15,798·002	0·000
2	6334·4279	15,782·381	15·621
3	6351·8618	15,739·064	58·938
4	6382·9914	15,662·306	135·696
5	6402·2460	15,615·202	182·800
6	6421·7108	15,567·871	230·131
7	6444·7118	15,512·310	285·692
8	6506·5279	15,364·935	433·067
9	6532·8824	15,302·951	495·051
10	6598·9529	15,149·735	648·267
11	6652·0925	15,028·714	769·288
12	6666·8967	14,995·342	802·660
13	6678·2764	14,969·790	828·212
14	6717·0428	14,883·395	914·607
15	6929·4672	14,427·144	1370·858
16	7024·0500	14,232·876	1565·126
17	7032·4128	14,215·950	1582·052
18	7051·2937	14,177·885	1620·117
19	7059·1079	14,162·191	1635·811
20	7173·9380	13,935·504	1862·498
21	7245·1665	13,798·503	1999·499
22	7438·8981	13,439·150	2358·852
23	7472·4383	13,378·828	2419·174
24	7488·8712	13,349·471	2448·531
25	7535·7739	13,266·384	2531·618
26	7544·0439	13,251·841	2546·161
27	7724·6281	12,942·045	2855·957
28	7839·0550	12,753·131	3044·871
29	7927·1172	12,611·457	3186·545
30	7936·9946	12,595·763	3202·239
31	7943·1805	12,585·954	3212·048
32	8082·4576	12,369·073	3428·929
33	8118·5495	12,314·085	3483·917
34	8128·9077	12,298·394	3499·608
35	8136·4061	12,287·060	3510·942
36	8248·6812	12,119·819	3678·183
37	8259·3795	12,104·120	3693·882
38	8266·0788	12,094·310	3703·692
39	8267·1166	12,092·792	3705·210

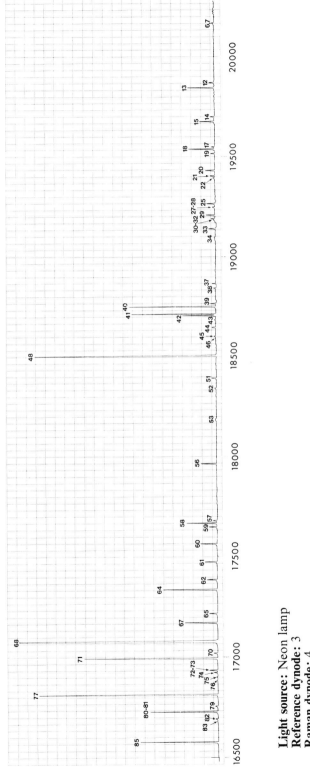

Light source: Neon lamp
Reference dynode: 3
Raman dynode: 4
Spectral slit width: 1 cm^{-1}
Scan speed: 250 cm^{-1}/min
Slit length: 10 cm
Sensitivity: 10
Single/Double slits: Single
Pen response: 0·1 sec

TABLE 3. Calibration of Raman spectrometers for argon
ion laser excitation

	Wavelength (Å)	Frequency (cm^{-1}_{vac})	Apparent Raman shift (cm^{-1}_{vac})
1	4884·9170	20,465·461	21·187
2	4892·1007	20,435·410	51·239
3	4928·2410	20,285·553	201·096
4	4939·0457	20,241·176	245·472
5	4944·9899	20,216·845	269·803
6	4957·0335	20,167·727	318·921
7	4957·1230	20,167·363	319·285
8	4994·9130	20,014·785	471·863
9	5005·1587	19,973·815	512·834
10	5011·0000	19,950·532	536·117
11	5022·8640	19,903·409	583·239
12	5031·3504	19,869·838	616·810
13	5037·7512	19,844·593	642·056
14	5074·2007	19,702·045	784·603
15	5080·3852	19,678·061	808·587
16	5104·7011	19,584·327	902·321
17	5113·6724	19,549·970	936·679
18	5116·5032	19,539·153	947·495
19	5122·2565	19,517·207	969·441
20	5144·9384	19,431·165	1055·483
21	5151·9610	19,404·679	1081·969
22	5154·4271	19,395·395	1091·253
23	5156·6672	19,386·970	1099·679
24	5158·9018	19,378·572	1108·076
25	5188·6122	19,267·610	1219·038
26	5191·3223	19,257·552	1229·096
27	5193·1302	19,250·848	1235·801
28	5193·2227	19,250·505	1236·143
29	5203·8962	19,211·022	1275·627
30	5208·8648	19,192·697	1293·952
31	5210·5672	19,186·426	1300·222
32	5214·3389	19,172·548	1314·100
33	5222·3517	19,143·132	1343·517
34	5234·0271	19,100·430	1386·218
35	5274·0393	18,955·524	1531·125
36	5280·0853	18,933·819	1552·829
37	5298·1891	18,869·123	1617·525
38	5304·7580	18,845·758	1640·891
39	5326·3968	18,769·197	1717·452
40	5330·7775	18,753·773	1732·876
41	5341·0938	18,717·550	1769·098
42	5343·2834	18,709·880	1776·768
43	5349·2038	18,689·173	1797·476
44	5360·0121	18,651·487	1835·161
45	5372·3110	18,608·789	1877·860

	Wavelength (Å)	Frequency (cm_{vac}^{-1})	Apparent Raman shift (cm_{vac}^{-1})
46	5374·9774	18,599·557	1887·091
47	5383·2503	18,570·974	1915·674
48	5400·5616	18,511·446	1975·202
49	5412·6493	18,470·106	2016·542
50	5418·5584	18,449·964	2036·684
51	5433·6513	18,398·717	2087·932
52	5448·5091	18,348·545	2138·103
53	5494·4158	18,195·242	2291·407
54	5533·6788	18,066·143	2420·506
55	5538·6510	18,049·925	2436·724
56	5562·7662	17,971·677	2514·971
57	5652·5664	17,686·171	2800·478
58	5656·6588	17,673·375	2813·273
59	5662·5489	17,654·992	2831·656
60	5689·8163	17,570·385	2916·264
61	5719·2248	17,480·038	3006·611
62	5748·2985	17,391·628	3095·020
63	5760·5885	17,354·524	3132·124
64	5764·4188	17,342·993	3143·656
65	5804·4496	17,223·387	3263·262
66	5811·4066	17,202·768	3283·880
67	5820·1558	17,176·908	3309·740
68	5852·4878	17,082·016	3404·633
69	5868·4183	17,035·645	3451·003
70	5872·8275	17,022·855	3463·793
71	5881·8950	16,996·613	3490·035
72	5902·4623	16,937·388	3549·260
73	5902·7835	16,936·467	3550·182
74	5906·4294	16,926·012	3560·636
75	5913·6327	16,905·395	3581·253
76	5918·9068	16,890·332	3596·317
77	5944·8342	16,816·668	3669·981
78	5961·6228	16,769·311	3717·338
79	5965·4710	16,758·493	3728·155
80	5974·6273	16,732·811	3753·838
81	5975·5340	16,730·272	3756·377
82	5987·9074	16,695·701	3790·948
83	5991·6532	16,685·263	3801·385
84	6000·9275	16,659·477	3827·172
85	6029·9971	16,579·165	3907·484
86	6046·1348	16,534·914	3951·735
87	6064·5359	16,484·744	4001·905

ROUTINE CALIBRATION

The neon emission spectrum is not the most convenient standard for routine use as the conditions required to record it are very different from those needed to record a Raman spectrum.

In the infrared region of the spectrum, the International Union of Pure and

Applied Chemistry (IUPAC) Commission on Molecular Structure and Spectroscopy recommended that polystyrene film and indene should be used as routine standards, and with this in mind the positions of the absorption maxima for these two compounds were measured interferometrically with an accuracy of 0.5 cm^{-1}.

Hendra and Loader[15] have considered the problem of routine calibration in Raman work and have concluded that indene provides an excellent standard in the majority of cases. The positions of the indene Raman emissions were located accurately by comparison with the neon emission spectrum, enabling vibrations which were active in both the infrared and Raman to be identified. A vibration was considered to be active in both if a frequency measured in the Raman coincided with one in the IUPAC list of infrared frequencies. In no case was there a discrepancy of more than 0.6 cm^{-1} between the absorption and Raman frequency values, taking into account all the experimental errors.

The 'correct' frequencies of these vibrations are those listed in the IUPAC tables and it is these values that are given in Table 4. Where no IUPAC value is available the frequency value measured from the Raman spectrum is given.

There is a need for another standard to cover the region from $\Delta\nu = 1700$ to 2800 cm^{-1} as none of the infrared absorptions in this region are strongly active in the Raman.

Indene, which is colourless when pure, normally contains impurities and before use as a calibration substance it should be vacuum distilled and sealed into Pyrex capillaries and ampoules. Both are useful so that the chart can be calibrated with as little disturbance as possible to the sampling system and the settings of the spectrometer.

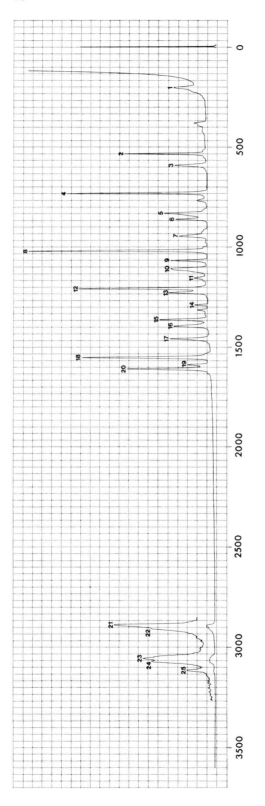

Compound: Indene
Phase: Liquid
Source: BDH
Purity: Reagent (vacuum distilled)
Sample cell: Sealed capillary
Reference dynode: 3
Raman dynode: 4
Spectral slit width: 4·6 cm^{-1}
Scan speed: 250 cm^{-1}/min
Slit length: 10 cm
Sensitivity: 40
Single/Double slits: Single
Pen response: 0·5 sec
Laser: He–Ne
Power: 40 mW

TABLE 4. Routine calibration of the laser Raman
spectrometer with indene

Line	Best (cm^{-1})	Source of value[a]	Calibration lines[b]
1	205·0 ± 2	N	
2	533·7 ± 0·5	N	←
3	593·0 ± 2	N	
4	730·1 ± 0·2	I	←
5	830·5 ± 0·2	I	
6	861·3 ± 0·2	I	
7	947·2 ± 0·3	I	
8	1018·6 ± 0·2	I	←
9	1067·9 ± 0·2	I	←
10	1108·9 ± 1	N	
11	1154·5 ± 0·5	I	
12	1205·2 ± 0·2	I	←
13	1226·2 ± 0·2	I	←
14	1287·8 ± 0·2	I	
15	1361·3 ± 0·4	I	←
16	1393·2 ± 1	I	←
17	1457·8 ± 0·5	I	←
18	1553·3 ± 0·5	I	←
19	1589·8 ± 1	N	
20	1609·6 ± 0·2	I	←
21	2892·2 ± 1	N	←
22	2901·2 ± 1	N	←
23	3054·7 ± 1	N	←
24	3068·5 ± 2·5	I	←
25	3112·7 ± 0·5	N	←

[a] I: IUPAC value for the position of the line obtained from interferometric measurements in the infrared.
N: Raman lines measured relative to the neon emission spectrum on the Cary 81 spectrometer.
[b] Those lines indicated with an horizontal arrow are recommended for calibration purposes.

4
Resolution checks

The resolution of the spectrometer is an item of information which is always given by the manufacturers and is used as one of the criteria for comparing different instruments. Apart from this it can give an indication of the internal conditions of the spectrometer, the operator being able to reproduce the manufacturer's spectrum only if the instrument is in the peak of condition.

There are two peaks that are frequently quoted, these being the 459 cm^{-1} line of carbon tetrachloride and the 471 cm^{-1} line of boron trichloride. Both these lines are multiplets since the compounds contain two isotopes of chlorine, ^{35}Cl and ^{37}Cl.

These different isotopes give rise to five different types of carbon tetrachloride molecule, each of which has a different symmetric stretching frequency. These differences are small, so that the splittings of the peak are small. The isotopic modifications and vibrational frequencies of carbon tetrachloride are shown in Table 5.

TABLE 5

Isotopic modification	Percentage	Frequency shift (cm^{-1})
C $^{37}Cl_4$	0·4	Not observed
C ^{35}Cl $^{37}Cl_3$	4·7	452·0 ± 0·2
C $^{35}Cl_2$ $^{37}Cl_2$	21·1	456·4 ± 0·2
C $^{35}Cl_3$ ^{37}Cl	42·2	459·4 ± 0·2
C $^{35}Cl_4$	31·6	462·4 ± 0·2

For boron trichloride there will be four lines and the details are shown in Table 6.

TABLE 6

Isotopic modification	Percentage	Frequency shift (cm^{-1})
B $^{37}Cl_3$	1·6	Not observed
B ^{35}Cl $^{37}Cl_2$	14·0	464·8 ± 0·2
B $^{35}Cl_2$ ^{37}Cl	42·2	469·2 ± 0·2
B $^{35}Cl_3$	42·2	473·7 ± 0·2

To produce spectra at this resolution it is necessary to use the least possible slit width consistent with a reasonable peak height at the maximum obtainable sensitivity of the instrument. As a result of using high sensitivity a long pen response time is required to maintain a good signal-to-noise ratio. The narrow slits and long response

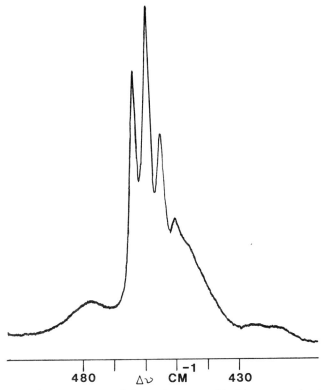

FIG. 14. Multiplet structure of the carbon tetrachloride symmetric stretching vibration. [Instrument: Coderg PH1 No. 8; phase: 30% in cyclohexane; sample cell: capillary; spectral slit width 0·8 cm⁻¹; scan speed: 1·5 cm⁻¹/min; laser: OIP He-Ne; power: 150 mW.]

(by courtesy of Dr. H. J. Clase)

time in turn mean that the spectrum has to be scanned very slowly, a scan speed of around 1 cm⁻¹/min generally being used. The values for the remaining variables can be obtained from the spectra shown above.

ARGON ION LASER PLASMA EMISSION

The discharge from the plasma in a high power argon ion laser is very intense and it can give rise to spurious 'Raman' lines when a weak spectrum is being recorded. Passing the laser beam through a small aperture and an interference filter is not sufficient to guarantee the complete removal of these emission lines from the light entering the instrument, and they may appear on the recorded spectrum if the instrument is being used at very high sensitivity.

It is therefore important that the positions and relative intensities of all the plasma emission lines which occur in the Raman spectral region should be known.

The plasma emission spectrum was recorded on the Cary 81 spectrometer and was produced by a Spectra Physics Model 140 argon ion laser. The frequencies of the emission lines were read from the instrument's wavenumber scale, which had been previously checked against the neon emission spectrum, and are accurate to ± 2 cm^{-1}.

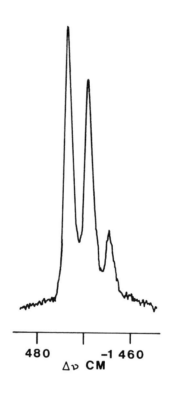

480   -1 460
$\Delta\nu$ CM

FIG. 15. Multiplet structure of the boron trichloride symmetric stretching vibration. [Source: BDH; purity: reagent; sample cell: sealed capillary; dynode: reference 2; Raman: 4; spectral slit width: 0·4 cm^{-1}; scan speed: 1·25 cm^{-1}/min; slit length: 2·5 cm; sensitivity: 3000; double slits; pen response: 20 sec; laser: He-Ne; power: 40 mW.]

The peak heights were measured and serve to give an indication of those lines which will most readily interfere in the Raman spectrum. It should, however, be borne in mind that these values are instrument dependent and give only a rough guide to their true intensities.

The argon ion laser can be tuned to operate on several wavelengths and each of these lines can be used as a Raman source. The power emitted by the laser at each of these wavelengths is shown in Table 7.

The apparent Raman shift of a plasma emission line will depend upon the laser line being used to excite the Raman spectrum at the time. In Table 8 the wavelengths shown are those calculated from the observed frequencies of the emission lines. The measured peak heights and apparent Raman shifts of the emission lines, with respect to all the possible exciting lines, are also shown.

TABLE 7. Wavelengths, frequencies and measured
power output per line for Spectra-Physics Model
140 argon ion laser No. 153

Wavelength in standard air (Å)	Frequency (cm_{vac}^{-1})	Power output (mW)
5145·27	19,429·91	1900
5017·17	19,926·00	250
4965·09	20,135·00	550
4879·86	20,486·68	1900
4764·88	20,981·02	550
4726·89	21,149·64	70
4657·95	21,462·66	20
4579·36	21,830·99	105

TABLE 8. Possible spurious lines due to emission from the laser plasma when an argon ion laser is being used as the Raman exciting source

Line	Frequency of emission line in vacuo (cm⁻¹)	Wavelength in standard air (Å)	Peak height	Apparent Raman displacement of the emission lines in wavenumbers (cm⁻¹) from the laser lines							
				4579 Å	4658 Å	4727 Å	4765 Å	4880 Å	4965 Å	5017 Å	5145 Å
1	21,995	4545·2	350								
2	21,903	4564·3	23								
3	21,831	4579·4	380	0							
4	21,783	4589·4	530	48							
5	21,739	4598·7	15	92							
6	21,688	4609·6	819	143							
7	21,557	4637·6	74	274							
8	21,463	4657·9	366	368	0						
9	21,150	4726·8	500	681	313						
10	21,126	4732·2	23	705	337	0					
11	21,108	4736·2	800	723	355	24					
12	20,981	4764·9	470	850	482	42					
13	20,803	4805·7	1150	1028	660	169	0				
14	20,623	4847·6	840	1208	840	347	178				
15	20,547	4865·5	40	1284	916	527	358				
16	20,487	4879·8	1600	1344	976	603	434				
17	20,450	4888·6	90	1381	1013	663	494	0			
18	20,384	4904·4	60	1447	1079	700	531	37			
19	20,267	4932·8	460	1564	1196	766	597	103			
20	20,226	4942·8	10	1605	1237	883	714	220			
21	20,135	4965·1	530	1696	1328	924	755	261	0		
22	20,107	4972·0	270	1724	1356	1015	846	352	28		
23	19,959	5008·9	830	1872	1504	1191	1022	528	176		

24	19,926	5017·2	330	1905	1537	1224	1055	561	209	0	
25	19,750	5061·9	790	2081	1713	1400	1231	737	385	176	
26	19,639	5090·5	5	2192	1824	1511	1342	848	496	287	
27	19,444	5141·5	27	2387	2019	1706	1537	1043	691	482	
28	19,430	5145·2	95	2401	2033	1720	1551	1057	705	496	0
29	19,364	5162·8	7	2467	2099	1786	1617	1123	771	562	66
30	19,353	5165·7	21	2478	2110	1797	1628	1134	782	573	77
31	19,313	5176·4	26	2518	2150	1837	1668	1174	822	613	117
32	19,269	5188·2	3	2562	2194	1881	1712	1218	866	657	161
33	19,163	5216·9	8	2668	2300	1987	1818	1324	972	763	267
34	18,909	5287·0	75	2922	2554	2241	2072	1578	1226	1017	521
35	18,842	5305·8	4	2989	2621	2308	2139	1645	1293	1084	588
36	18,521	5397·8	4	3310	2942	2629	2460	1966	1614	1405	909
37	18,503	5403·0	3	3328	2960	2647	2478	1984	1632	1423	927
38	18,487	5407·7	4	3344	2976	2663	2494	2000	1648	1439	943
39	18,336	5452·2	3	3495	3127	2814	2645	2151	1799	1590	1094
40	18,329	5454·3	6	3502	3134	2821	2652	2158	1806	1597	1101
41	18,190	5496·0	7	3641	3273	2960	2791	2297	1945	1736	1240
42	18,183	5498·1	3	3648	3280	2967	2798	2304	1952	1743	1247
43	18,175	5500·5	3	3656	3288	2975	2806	2312	1960	1751	1255
44	17,984	5559·0	10	3847	3479	3166	2997	2503	2151	1942	1446
45	17,940	5572·6	4	3891	3523	3210	3041	2547	2195	1986	1490
46	17,830	5607·0	12	4001	3633	3320	3151	2657	2305	2096	1600
47	17,692	5650·7	8	4139	3771	3458	3289	2795	2443	2234	1738
48	17,680	5654·5	3	4151	3783	3470	3301	2807	2455	2246	1750
49	17,462	5725·1	3	4369	4001	3688	3519	3025	2673	2464	1968
50	17,417	5739·9	3	4414	4046	3733	3564	3070	2718	2509	2013
51	17,318	5772·7	7	4513	4145	3832	3663	3169	2817	2608	2112
52	17,275	5787·1	3	4556	4188	3875	3706	3212	2860	2651	2155
53	17,197	5813·4	5	4634	4266	3953	3784	3290	2938	2729	2233
54	17,106	5844·3	4	4725	4357	4044	3875	3381	3029	2820	2324
55	17,058	5860·7	3	4773	4405	4092	3923	3429	3077	2868	2372
56	16,993	5883·1	5	4838	4470	4157	3988	3494	3142	2933	2437
57	16,977	5888·7	11	4854	4486	4173	4004	3510	3158	2949	2453
58	16,909	5912·4	19	4922	4554	4241	4072	3578	3226	3017	2521

TABLE 8 (*Contd.*)

Line	Frequency of emission line in vacuo (cm⁻¹)	Wavelength in standard air (Å)	Peak height	Apparent Raman displacement of the emission lines in wavenumbers (cm⁻¹) from the laser lines							
				4579 Å	4658 Å	4727 Å	4765 Å	4880 Å	4965 Å	5017 Å	5145 Å
59	16,861	5929·2	6	4970	4602	4289	4120	3626	3274	3065	2569
60	16,701	5986·0	6	5130	4762	4449	4280	3786	3434	3225	2729
61	16,691	5989·6	6	5140	4772	4459	4290	3796	3444	3235	2739
62	16,572	6032·6	48	5259	4891	4578	4409	3915	3563	3354	2858
63	16,541	6043·9	16	5290	4922	4609	4440	3946	3594	3385	2889
64	16,515	6053·4	8	5316	4948	4635	4466	3972	3620	3411	2915
65	16,498	6059·7	15	5333	4965	4652	4483	3989	3637	3428	2932
66	16,449	6077·7	6	5382	5014	4701	4532	4038	3686	3477	2981
67	16,391	6099·2	5	5440	5072	4759	4590	4096	3744	3535	3039
68	16,378	6104·1	44	5453	5085	4772	4603	4109	3757	3548	3052
69	16,348	6115·3	1020	5483	5115	4802	4633	4139	3787	3578	3082
70	16,324	6124·3	47	5507	5139	4826	4657	4163	3811	3602	3106
71	16,284	6139·3	51	5547	5179	4866	4697	4203	3851	3642	3146
72	16,266	6146·1	8	5565	5197	4884	4715	4221	3869	3660	3164
73	16,240	6155·9	4	5591	5223	4910	4741	4247	3895	3686	3190
74	16,196	6172·7	950	5635	5267	4954	4785	4291	3939	3730	3234
75	16,157	6187·6	12	5674	5306	4993	4824	4330	3978	3769	3273
76	16,119	6202·1	5	5712	5344	5031	4862	4368	4016	3807	3311
77	16,091	6212·9	9	5740	5372	5059	4890	4396	4044	3835	3339
78	16,082	6216·4	6	5749	5381	5068	4899	4405	4053	3844	3348
79	16,020	6240·5	17	5811	5443	5130	4961	4467	4115	3906	3410
80	16,013	6243·2	470	5818	5450	5137	4968	4474	4122	3913	3417
81	15,876	6297·1	9	5955	5587	5274	5105	4611	4259	4050	3554
82	15,848	6308·2	14	5983	5615	5302	5133	4639	4287	4078	3582

83	3624	4120	4329	4681	5175	5344	5657	6025	13	6325·0	15,806
84	3645	4141	4350	4702	5196	5365	5678	6046	10	6333·4	15,785
85	3683	4179	4388	4740	5234	5403	5716	6084	9	6348·7	15,747
86	3706	4202	4411	4763	5257	5426	5739	6107	7	6357·9	15,724
87	3725	4221	4430	4782	5276	5445	5758	6126	4	6365·6	15,705
88	3736	4232	4441	4793	5287	5456	5769	6137	9	6370·1	15,694
89	3773	4269	4478	4830	5324	5493	5806	6174	19	6385·2	15,657
90	3796	4292	4501	4853	5347	5516	5829	6197	7	6394·5	15,634
91	3802	4298	4507	4859	5353	5522	5835	6203	8	6397·0	15,628
92	3809	4305	4514	4866	5360	5529	5842	6210	104	6399·9	15,621
93	3819	4315	4524	4876	5370	5539	5852	6220	7	6404·0	15,611
94	3832	4328	4537	4889	5383	5552	5865	6233	5	6409·3	15,598
95	3851	4347	4556	4908	5402	5571	5884	6252	79	6417·1	15,579
96	3867	4363	4572	4924	5418	5587	5900	6268	5	6423·7	15,563
97	3887	4383	4592	4944	5438	5607	5920	6288	3	6432·0	15,543
98	3902	4398	4607	4959	5453	5622	5935	6303	19	6438·2	15,528
99	3913	4409	4618	4970	5464	5633	5946	6314	14	6442·8	15,517
100	3917	4413	4622	4974	5468	5637	5950	6318	11	6444·4	15,513
101	3977	4473	4682	5034	5528	5697	6010	6378	8	6469·4	15,453
102	3986	4482	4691	5043	5537	5706	6019	6387	6	6473·2	15,444
103	3993	4489	4698	5050	5544	5713	6026	6394	5	6476·2	15,437
104	4011	4507	4716	5068	5562	5731	6044	6412	480	6483·7	15,419
105	4052	4548	4757	5109	5603	5772	6085	6453	56	6501·0	15,378
106	4072	4568	4777	5129	5623	5792	6105	6473	10	6509·5	15,358
107	4141	4637	4846	5198	5692	5861	6174	6542	65	6538·8	15,289
108	4199	4695	4904	5256	5750	5919	6232	6600	15	6563·7	15,231
109	4277	4773	4982	5334	5828	5997	6310	6678	3	6597·5	15,153
110	4295	4791	5000	5352	5846	6015	6328	6696	28	6605·4	15,135
111	4317	4813	5022	5374	5868	6037	6350	6718	19	6615·0	15,113
112	4332	4828	5037	5389	5883	6052	6365	6733	24	6621·6	15,098
113	4371	4867	5076	5428	5922	6091	6404	6772	2800	6638·7	15,059
114	4383	4879	5088	5440	5934	6103	6416	6784	5700	6644·0	15,047
115	4434	4930	5139	5491	5985	6154	6467	6835	37	6666·6	14,996

5

Factors affecting the intensity of the recorded spectrum

The newcomer to laser Raman spectroscopy is presented with a considerable number of unrelated facts which appear to have a substantial bearing on the type and quality of spectra obtained and on the prices of the various instruments.

It is the object of this section to explain some of them and remove a little of the mystique surrounding them.

1. RAMAN SCATTERING

The theory of the Raman effect[3] shows that the amount of Raman scattering from a particular set of molecules is directly proportional to the intensity of the incident light, to the fourth power of the frequency (Hz) of the laser line exciting the spectrum and to other parameters, mainly molecular, over which the experimentalist has but little control.

The Raman scattering is directly proportional to the intensity of the incident light – the higher the power that can be concentrated into a particular volume of sample, the more intense will be the recorded spectrum. Therefore very high power lasers are used to bring weak spectra, such as those from gases, up to a level where they can be recorded photo-electrically.

It must be remembered that a level will be reached at which the sample will decompose in the beam in a shorter time than it will take to record the spectrum. For some unstable species this level is reached very soon and care must be taken to ensure that the recorded spectrum is of the compound and not of its decomposition products.

The Raman scattering is proportional to the fourth power of the frequency of the laser line exciting the spectrum – this means that a red laser will be less efficient as a source for Raman spectroscopy than a green one.

As the frequency in wavenumbers is obtained by dividing the frequency in hertz by the velocity of light, the scattering is also proportional to the frequency in wavenumbers.

Changing from excitation of the spectrum by a 100 mW krypton ion laser operating at 14783 cm^{-1} (6471 Å) to a 100 mW argon ion laser operating at 20487 cm^{-1} (4880 Å) causes an increase of scattering efficiency of $\left(\dfrac{20487}{14783}\right)^4 = 3\cdot7$ times.

2. DIFFRACTION GRATING EFFICIENCY

The majority of diffraction gratings used for laser Raman spectroscopy are blazed at 5000 Å in the first order. The efficiency of a grating in a particular order is at a maximum at the blaze wavelength and falls on either side of that wavelength. For a grating operating in the first order the efficiency has fallen to about one half at two thirds and twice the blaze wavelength. The general shape of the curve is indicated in Fig. 16.

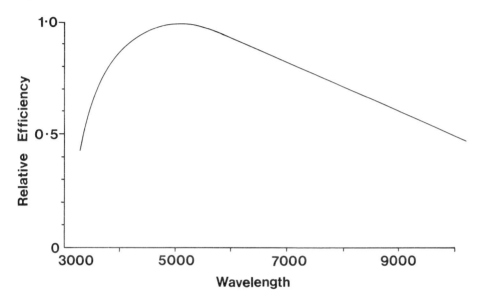

FIG. 16. Relative efficiency of a diffraction grating

This type of grating would be fitted in a spectrometer where a number of different laser lines including those of the argon ion laser were going to be used. For an instrument which was going to be used exclusively with a helium-neon laser, gratings blazed at 6500 Å would most probably be favoured as they have a higher efficiency in the red and would make the detection of C–H stretching vibrations easier.

3. PHOTOMULTIPLIER TUBES

The Raman light emerging from the exit slit of the spectrometer is generally detected by a photomultiplier tube having an S-20 type spectral response. The general shape of a standard S-20 response is shown in Fig. 17 together with the spectral regions used when the Raman spectrum is excited by one of the four most popular laser lines.

It is immediately obvious that the efficiency of the photomultiplier tube falls very rapidly at long wavelengths and it is usual for the spectrometer manufacturers to try to select a tube with a better than average red response.

A failure to appreciate the magnitude of the decrease in the efficiency of the photomultiplier tube coupled with the decrease in grating efficiency over the Raman range when the spectrum is excited by a helium-neon laser has led to a false but widely held belief that —CH stretching vibrations are weak in the Raman. A decrease in the spectral response of a spectrometer by as much as 100 times is the real cause of the apparent weakness.

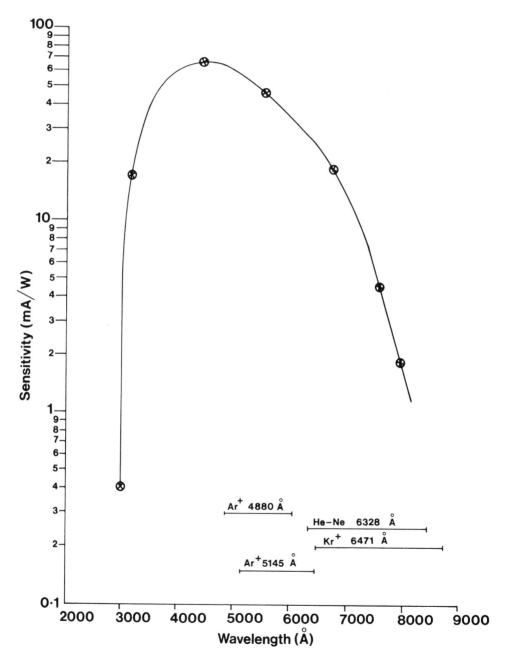

FIG. 17. Typical S-20 photomultiplier response

4. SPECTRAL SLIT WIDTH

Under ideal conditions the spectral slit width of the Raman spectrometer should remain constant during the scan of the spectrum, as is usual with modern infrared spectrometers.

In order to achieve this condition the actual slit width of the spectrometer has to be continually varied throughout the scan. At present only the Jarrell-Ash 25-300 spectrometer has this facility and all the others either set the actual slit width in

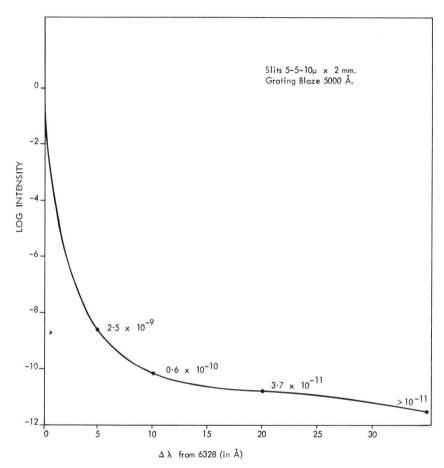

FIG. 18. Discrimination curve for the Jarrell-Ash 25–300 spectrometer

microns (e.g. Spex) or else set the spectral slit width at some predetermined point (e.g. Cary. The slit setting dial reads spectral slit width at 6328 Å and a correction table is supplied with the instrument).

5. DISCRIMINATION

The discrimination of the spectrometer is a measure of the stray light in the monochromator. Due to the low intensity of the Raman light it is essential that the level of stray light is as low as possible. For this reason a double monochromator is always preferable to a single one because the stray light intensity in the former is reduced to about the square root of that in the latter.

The discrimination is determined by passing a laser beam straight through the entrance slit of the monochromator and measuring the intensity of the radiation reaching the detector when the monochromator is set first at the laser frequency and then at a series of frequencies at small displacements from the laser line. The results are plotted graphically as shown in Fig. 3.

In the case of a perfect monochromator, no radiation would be detected at any

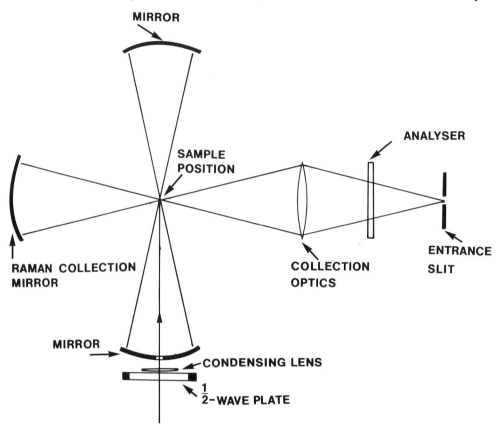

FIG. 19. 90° sample illumination system

frequency other than that corresponding to the laser emission, but due to imperfections light does reach the detector at other frequencies. In a good double monochromator an intensity ratio of about 10^{-10} should have been reached by the time the displacement from the laser line is 50 cm^{-1}.

6. SAMPLE ILLUMINATING METHODS

In classical Raman spectroscopy using a mercury-arc lamp the sample was totally surrounded by the lamp and the Raman radiation was observed at 90° to the illumination direction, i.e. along the axis of the lamp.

With the advent of the laser as a source, there came the possibility of illuminating the sample from a variety of directions. By far the most popular of these is the 90° method, however. In this method the laser is directed at the sample from the side, top or bottom, and the cell is usually surrounded by mirrors to reflect as much light as possible into the slit of the spectrometer.

The sample may also be illuminated and viewed along the same direction, i.e. 0° illumination. This can, however, only be done on an instrument with very good discrimination such as the Spex or Jarrell-Ash.

The Cary spectrometer is unique in that it uses 180° illumination in conjunction with an image slicer.

The laser beam is directed downwards through the one millimetre face prism, which is stuck to the rear of the hemispherical lens, and illuminates the sample contained in the tube pressed against the lens. The Raman light is collected by the hemispherical lens over a large angle and passed through the image slicer to the slits of the mono-chromator. The image slicer divides the circular image of the sample tube into two sets of ten strips, places them end to end and illuminates the two 10 cm high slits.

This sampling method has several advantages over the others for general use:

(i) The samples need not be transparent to the laser radiation.
(ii) Many sizes and shapes of sample can be accommodated provided that they can be brought into contact with the hemispherical lens.
(iii) The instrument has a very high light-gathering capacity.

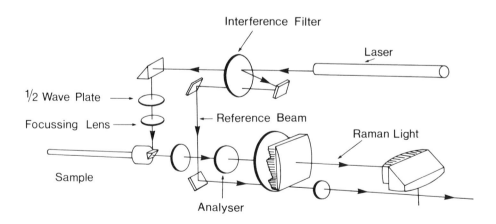

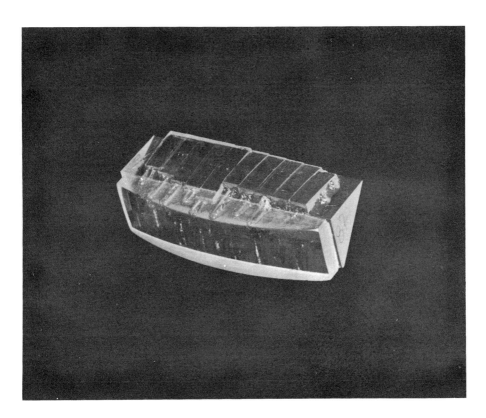

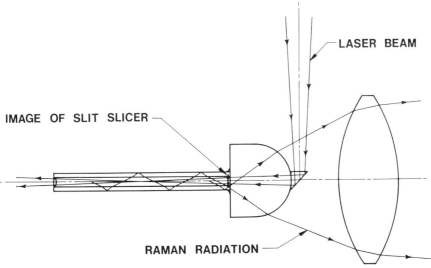

FIG. 20. Cary pre-slit optics and image slicer

(by courtesy of Cary Instruments)

6

Chemical applications of laser Raman spectroscopy

Infrared spectroscopy is used most frequently for the identification of unknown compounds by comparison of their spectra with sets of standard spectra, particular attention being paid to strong absorptions in the so-called finger-printing regions.

As yet, Raman has not been used extensively for fingerprinting organic compounds but the reason for this has been the experimental difficulties of the technique. However, in 1943 Kohlrausch produced a book in which he listed details of the Raman spectra of several hundred organic compounds (see Bibliography). This book is of considerable value to the Raman spectroscopist but it is unfortunately out of print at the moment.

Although they have been mentioned in various places in the book the reasons for using Raman spectroscopy as well as infrared will be listed here.

Theoretical reasons:

1. Raman spectroscopy constitutes one half of vibrational spectroscopy.
2. Totally symmetric vibrations are only active in the Raman.
3. For molecules with a centre of symmetry, the mutual exclusion principle applies and vibrations allowed in the infrared are forbidden in the Raman and vice versa.
4. Polarisation measurements carried out on pure liquids and solutions will identify totally symmetric vibrations.

Practical reasons:

1. Glass is almost completely transparent in the Raman so spectra can be readily obtained from samples sealed into tubes and capillaries. These tubes can be kept and the sample preserved for future reference. Glass does have some weak emissions but they generally appear as small bumps on the background and do not cause any trouble.
2. The sample can be any convenient shape and need not be transparent to the laser radiation. In addition, only a small quantity of sample is necessary and it is generally uncontaminated by mulling agents or solvents.
3. The whole of the vibrational spectrum from less than $40 \, cm^{-1}$ to $4000 \, cm^{-1}$ can be obtained in one run on one instrument.
4. Water is a very good solvent for Raman spectroscopy.
5. The widths of Raman lines are generally less than those of infrared bands.

Hence the lines due to organic solvents in solution studies interfere less in the Raman than they do in the infrared.

Chemical reasons:

Several spectroscopists, in particular Dr. H. Sloane of Cary Instruments, have been investigating the Raman spectra of a wide range of organic compounds and comparing them with the infrared spectra with a view to assessing the potentialities of laser Raman spectroscopy as a routine analytical tool.

Some of their preliminary conclusions will be discussed below.

1. Symmetric and pseudo-symmetric vibrations are strong in the Raman and weak or absent in the infrared. Vibrations such as C=C, C≡C, S—S, N=N, etc., are therefore readily observed.
2. The nitrile group frequency is of variable intensity in the infrared, and is very weak when there is a strongly electronegative substituent adjacent to it. The nitrile stretching vibration is always strong in the Raman.
3. C=S and S—H are relatively strong in the Raman, whereas the latter is virtually undetectable in the infrared.
4. O—H vibrations are weak in the Raman compared to C—H and N—H vibrations. In molecules containing N—H and C—H groups as well as O—H, the former two groups can be investigated without the tremendous interference from the latter which is experienced in the infrared.
5. The symmetric ring stretching vibration of cyclic compounds gives rise to a strong polarised line. The position of the line is characteristic of the type and size of ring present in the compound.
6. Aromatic compounds have particularly strong spectra and the following conclusions have been drawn from an investigation of more than thirty of them.

For all aromatics, regardless of substitution, there is a strong band at 1600 ± 30 cm^{-1} (*cf.* spectrum of benzene in Appendix C) Monosubstituted compounds have a very strong symmetric ring stretching vibration at about 1000 cm^{-1}, a strong in-plane hydrogen bending vibration at about 1025 cm^{-1}, and a weak depolarised in-plane bending vibration at about 615 cm^{-1}.

Meta and 1,3,5-substituted compounds have the line at about 1000 cm^{-1} but neither the line at 1025 cm^{-1} nor the one at 615 cm^{-1}. *Ortho*-substituted compounds have a line at about 1037 cm^{-1} but nothing at 1000 cm^{-1}. *Para*-substituted compounds have no lines in the region 1000 cm^{-1}, unless it is fortuitously from other groups, but they do have a weak line at 640 cm^{-1}.

These conclusions have already proved to be of considerable value and taken together with the usual infrared criteria of aromaticity can provide strong evidence for the presence of a substituted aromatic compound.

The above points are the first fruits of the new study of group frequencies in the Raman and it is to be expected that very many more will follow.

Quantitative analytical reasons:

The intensity of a Raman line is directly proportional to the amount of compound giving rise to the line present in the area sampled by the spectrometer. The amount of a compound present in a mixture can readily be estimated by measuring the relative intensities of lines due to each of the components and comparing the ratios of the intensities of the lines with calibration curves obtained from measurements made on mixtures of known composition.

Eight-component mixtures of benzene and isopropyl benzenes have been studied by Nicholson[16] who used a Cary 81 spectrometer and a mercury arc as the exciting source. He studied synthetic mixtures and found that the average deviation between the measured and known concentrations of the components was about 1%. The sensitivity of the technique was good as he was able to detect a concentration of 0·001% benzene in carbon tetrachloride.

The factors affecting the use of Raman spectroscopy as an analytical tool have been discussed in detail by Tunnicliff and Jones[17].

References

1. A. Smekal, *Naturwissenschaften* **11,** 873 (1923).
2. C. V. Raman and K. S. Krishnan, *Nature* **121,** 501 (1928).
3. C. V. Raman and K. S. Krishnan, *Proc. Roy. Soc. (London)* **122a,** 23 (1928).
4. H. L. Welsh, M. F. Crawford, T. R. Thomas, and G. R. Love, *Can. J. Phys.*, **30,** 577 (1952).
5. S. P. S. Porto and D. L. Wood, *J. Opt. Soc. Am.* **52,** 251 (1962).
6. B. P. Stoicheff, *Tenth Intern. Spectr. Colloq., University of Maryland, June 1962*, Spartan Books, Washington, 1963.
7. J. J. Barrett and N. I. Adams, *J. Opt. Soc Am.* **58,** 311 (1968).
8. W. F. Murphy, M. V. Evans, and P. Bender, *J. Chem. Phys.* **47,** 1836 (1967).
9. J. R. Allkins and E. R. Lippincott. *Spectrochim. Acta* **25A,** 761 (1969).
10. P. J. Hendra and E. J. Loader, *Nature* **216,** 789 (1967).
11. P. J. Hendra and J. R. Mackenzie, *Chem. Commun.* 760 (1968).
12. G. L. Carlson, *Spectrochim. Acta* **25A,** 1519 (1968).
13. K. Burns, K. B. Adams, and J. Longwell, *J. Opt. Soc. Am.* **40,** 339 (1950).
14. B. Edlén, *J. Opt. Soc. Am.*, **43,** 339 (1953).
15. P. J. Hendra and E. J. Loader, *Chem. & Ind.* 718 (1968).
16. D. E. Nicholson, *Anal. Chem.* **32,** 1634 (1960).
17. D. D. Tunnicliff and A. C. Jones, *Spectrochim. Acta* **18,** 579 (1962).

Bibliographical reviews

As yet no book devoted entirely to laser Raman spectroscopy has been published so the books listed below deal in the main only with 'classical' Raman spectroscopy.

C. N. Banwell, *Fundamentals of Molecular Spectroscopy*, McGraw-Hill, London, 1966.

This book is exactly what the title suggests and gives a very good introduction to the theory of Raman spectroscopy. There are small sections on structure determination and on instrumentation.

J. Brandmüller and H. Moser, *Einführung in die Ramanspektroskopie*, Dietrich Steinkopff Verlag, Darmstadt, 1962.

A complete review of all aspects of Raman spectroscopy up to 1962. It includes large sections on experimental techniques, instrumentation and theory.

N. B. Colthup, L. H. Daly, and S. E. Wiberley, *Introduction to Infrared and Raman spectroscopy*, Academic Press, 1964.

Although mainly concerned with infrared spectroscopy, the sections in this book on Raman are informative and well written. It contains sections on vibrational analysis and on the calculation of thermodynamic functions as well as vast quantities of information on molecular group frequencies.

G. Herzberg, *Infrared and Raman spectra*, Van Nostrand, New York, 1945.

First published in 1945 and reprinted twelve times, this book is the vibrational spectrocopists' bible.

It contains a complete theoretical treatment of the 'classical' Raman effect, vibrational analysis and many other closely related topics as well as a large number of experimental results.

K. W. F. Kohlrausch, *Ramanspektren*, Akademische Verlagsgesellschaft, Becker and Erler Kom.-Ges, Leipzig, 1943.

Although this book was published in 1943 it is still of immense value as it contains details of the spectra of several hundred organic compounds.

H. A. Szymanski, Editor, *Raman Spectroscopy*, Plenum Press, New York, 1967.

Eight eminent research workers produced the specialised articles that make up this excellent volume. It is the only book to mention laser Raman and 'non-classical' effects. Koningstein describes the early work on laser Raman spectroscopy and Behringer discusses observed resonance Raman spectra. This is a valuable book for the research worker and interesting reading for the general student who can leave out the more difficult theoretical sections.

S. Walker and H. Straw, *Spectroscopy*, Vol. 2, Chapman and Hall, London, 1962.

This book has a large chapter on Raman spectroscopy and is a good introductory text. Experimental and applications sections are included and the correlation between infrared and Raman is discussed.

D. H. Whiffen, *Spectroscopy*, Longmans, London, 1966.

This small and well-written book provides a basic introduction to the majority of spectroscopy. Although the chapter on Raman only comprises six pages they are well worth reading.

It is to be confidently expected that several new books on Raman will appear in the very near future. Among them will be:

P. J. Hendra and T. R. Gilson, *Laser Raman Spectroscopy*, London, Wiley.

R. E. Hester and J. H. R. Clarke, *Raman Spectroscopy*, W. A. Benjamin Inc., New York.

Appendix A

OTHER COMMERCIAL INSTRUMENTS

New Raman spectrometers are appearing regularly and it has been decided to include a short appendix outlining the main details so that any prospective buyer would be aware of as full a range of instruments as possible. In general, they have not been used in British chemical laboratories over a long enough period for there to have been non-standard sampling systems developed, and thus the only information that can be given is that supplied by the manufacturers.

The manufacturers ask that any prospective buyer should submit samples of the type in which they are particularly interested so that the performance of their instrument may be demonstrated.

CODERG ROUTINE MODEL

This spectrometer has been designed specifically for control and routine work and is thus complementary to the wide range of small infrared instruments available. The complete instrument weighs about 120 kg and occupies an area 85 cm × 75 cm.

It accepts the standard transfer plates that are used with the Coderg PH 1 spectrometer and it is powered by a Spectra Physics 124 15 mW helium-neon laser. The spectrum is scanned linearly with respect to wavenumber and recorded on a non-coupled recorder.

A double monochromator is used in conjuction with the same type of detector system and electronics that are employed in the PH 1.

HUET R 50

This small desk top instrument is manufactured by Société Genérale d'Optique 'Huet', 76 Boulevard de la Villette 75, Paris 19e France.

It is supplied with a Cie Genérale d'Électricité 5 mW helium-neon laser and uses 90° sample illumination. It has a single monochromator which has a resolution of 1 cm^{-1} at 6328 Å. The Raman light is detected by an S 20 response photomultiplier tube, a phase sensitive detector and a.c. amplification. The spectrum is recorded on a non-coupled strip chart recorder.

A binary computing attachment is available which gives a considerable improvement in the signal-to-noise ratio by the summation of repeated scans over the same spectral region. The system works because noise is generally a random effect and the summation of random effects leads to an almost zero value. The Raman emissions, however, always occur in the same place and are thus summed to give a greater line intensity. This method has already proved very useful in n.m.r. spectroscopy.

JARRELL-ASH 25–300 LASER RAMAN

This instrument is manufactured by the Jarrell-Ash Co., 590 Lincoln Street, Waltham, Mass. 02154, USA, and the United Kingdom agents are Metals Research Ltd., Melbourn, Royston, Herts.

Sample illumination can be from above, below or on axis, and the instrument will accept almost any commercially available c.w. laser. The double monochromator has a resolution of 0.25 cm^{-1} at 6328 Å and shows very good discrimination.

The slits are electrically controlled and this enables the spectral slit width to be set and retained constant throughout the complete scan of the spectrum.

An S 20 response photomultiplier tube in conjunction with photon counting is used in the detection system. Cooling the photomultiplier tube thermoelectrically to $-22°C$ results in a hundredfold improvement in the signal-to-noise ratio over an uncooled tube.

The spectrum is output on a strip recorder which is calibrated automatically by a marker every 10 cm^{-1}.

Although there are many of these instruments in use, none have been sold in Great Britain with the result that the author was unable to use one personally and include a more detailed consideration in Chapter 2.

JEOL JRS-01A

This instrument is produced by the Japan Electron Optics Laboratory Co. Ltd., New Tokyo Building, 3-2 Marunouchi, Chiyoda-ku, Tokyo, Japan. The British subsidiary company is JEOLCO (U.K.) Ltd., Shakespeare Road, Finchley, London N3.

This is the only instrument so far produced to be supplied with an argon ion laser as the standard source, but a 50 mW helium-neon laser can be supplied as a special accessory. The argon ion laser produces 250 mW in either the 4880 Å or 5145 Å line.

The instrument has a double monochromator with a resolution of 0.2 Å (approx. 0.8 cm^{-1}) at 4880 Å, and 90° sample illumination is used. The sample can be adjusted into the correct position by the use of a lamp situated behind the intermediate slit of the monochromator. The spectrum is scanned linearly by wavelength covering the range 2000 to 8000 Å (50,000 to 12,500 cm^{-1}).

Detection is either by photographic plate or by HTV R 292 photomultiplier tube in conjunction with photon counting. The spectrum is presented on a non-coupled strip recorder which is automatically calibrated with a marker every 5 Å.

PERKIN-ELMER LR3 LASER-EXCITED RAMAN SPECTROMETER

This instrument is similar in appearance to the LR-1 but it can be used with both helium-neon and argon ion lasers.

The spectrometer has a double pass single grating monochromator followed by a grating post monochromator which tracks the main monochromator and acts as a variable stray light filter. The use of the post monochromator reduces the stray light by a factor of a hundred compared with a simple double pass single grating mono-chromator. The spectrum is scanned linearly with respect to wavenumber and output is on a pre-printed chart.

An S-20 response photomultiplier is used in conjunction with a phase-sensitive amplifier in the detection system.

2·5 ml and 25 μl liquid cells and solid sampling accessories are supplied.

The pre-printed chart can be put onto the Perkin-Elmer Model 521 (or after changing a gear, onto the Model 621) infrared spectrometer and the infrared spectrum recorded on the same chart as the Raman, thereby facilitating the comparison of the two spectra.

SPECTRA-PHYSICS MODEL 700 RAMAN SPECTROPHOTOMETER

This instrument is manufactured by Spectra-Physics, 1255 Terra Bella Avenue, Mountain View, California 94040, USA.

The standard source for the Model 700 is the Spectra-Physics Model 124A 15 mW helium-neon laser but it can accept the Model 141 argon ion laser.

A double monochromator is used with the two gratings mounted back to back. The entrance and exit slits can be adjusted to give spectral slit widths of 1, 2, 4, and 8 cm^{-1} at 6328 Å and the intermediate slit is continuously adjustable over the same range.

The slit selection wheel is linked to the electronics so that when the scan speed and slit selection is completed the time constants are automatically selected to match.

The detector is an FW-130 photomultiplier (S-20 response) used in conjunction with photon counting and d.c. amplification. When the signal is sufficiently strong the d.c. amplifier is automatically switched in instead of the photon counting system.

The spectrum output is on a flat bed recorder which can cover 125, 250, 500, 1000, 2000 or 4000 cm^{-1} full scale. It is linked to the monochromator by a master pulse clock which controls the stepping motors which drive the monochromator and the recorder.

Thermoelectric cooling is available as a field-fitted option but it is recommended that it be factory fitted if possible.

Sample holders for liquids, solids, single crystals and special applications are available.

Appendix B

LASERS

There are now a considerable number of lasers available which can be used for laser Raman spectroscopy. The main types used with photoelectric recording instruments are helium-neon, argon ion and krypton ion lasers, and the ones available from the various manufacturers are described.

For use as a source for laser Raman spectroscopy the laser must have the following characteristics:

1. The laser must have the highest possible amplitude stability. Any instability in the laser plasma appears as noise on the recorded spectrum, making detection of weak spectral lines difficult.

2. The laser should have as high a power as possible but at the same time it must be possible to reduce the power without introducing any instability so that highly reactive samples can be investigated without risk of destroying them.

3. The laser should have as many emission lines as possible over a wide spectral range to facilitate the study of coloured compounds.

COHERENT RADIATION LABORATORIES

932, East Meadow Drive, Palo Alto, California 94303, USA. British sales office: 23, Honey Way, Royston, Herts.

Model 52 ion lasers

The Model 52 has two options – either argon ion or krypton ion. The optical specifications will be considered separately later; both are driven by the same power supply and have approximately the same requirements.

The Model 52 laser is an extremely compact system for one delivering so much power. The laser head and power supply together weigh only 150 lb.

The laser is d.c. excited and the amplitude stability is better than 1% r.m.s. (10 Hz to 1 KHz), but can be improved further by the addition of the Model 235 amplitude stabiliser which at the expense of about 2% of the output power improves the amplitude stability to better than 0·2% r.m.s. (1 Hz to 1 KHz).

The output power can be varied between 35% and 100% of full power by means of the tube current control.

The various wavelengths are selected by a prism wavelength selector (Model 431) using the normal broad band mirrors.

Power requirements: 208/120 V 3-phase at about 35A per line.
Cooling: Flow through filtered tap water at 1·5 gallons per minute.

Power output of the Model 52 argon ion laser

Wavelength (Å)	Power (mW)
4579	50
4658	30
4727	30
4765	250
4880	700
4965	150
5017	150
5145	700

Power output of the Model 52 krypton ion laser

Wavelength (Å)	Power (mW)
4619	5
4680	5
4762	30
4825	20
5208	60
5308	60
5682	60
6471	150
6764	5

Model 53 ion lasers

These are basically similar to the Model 52 but they deliver over three times as much power. As before the optical specifications will be considered later. Both are driven by the same power supply and have approximately the same requirements.

The laser head and power supply weigh 330 lb. The laser is d.c. excited and the noise level is less than 1% r.m.s. (10 Hz to 2 MHz). Use of the Model 235 amplitude stabilizer will improve the noise level to better than 0·2% r.m.s. (1 Hz to 1KHz) at the expense of about 2% of the output power.

The various wavelengths are selected by a Model 431 prism wavelength selector, using the normal broad band mirrors.

Special mirrors are available which give very high power outputs from one line, i.e. 1 watt from the krypton 6471 Å line.

Power requirements: 208/120 V, 3-phase at about 50 A per line.
Cooling: Flow through filtered tap water at 2·5 gallons per minute.

Power output of the Model 53 argon ion laser

Wavelength (Å)	Power (mW)
4579	200
4658	50
4727	50
4765	600
4880	1500
4965	700
5017	500
5145	2500

Power output of the Model 53 krypton ion laser

Wavelength (Å)	Power (mW)
4619	10
4680	20
4762	75
4825	50
5208	200
5308	200
5682	200
6471	450
6764	10

Model 54 argon ion laser

This laser was announced at the beginning of May 1969, and is of a lower power than the Model 52 argon ion laser. It is rated at 500 mW (all-lines) as opposed to 3000 mW for the Model 52 operating in the same way.

All-line operation is where a wavelength selector is not used and the laser is lasing on all or most of its lines at once. In all-line operation the Model 52 lases on the six most powerful lines.

A wavelength selector will be available for the Model 54 and it will give in excess of 200 mW in the 4880 Å and 5145 Å lines.

FERRANTI LTD.

Laser Sales, Dunsinane Avenue, Dundee DD2 3PN, Scotland.

Type LR 50 helium-neon laser

This laser has a power of 50 mW at a wavelength of 6328 Å. In its standard form it is d.c. excited and has a self-generated current noise level of better than 7% r.m.s.

An auxiliary r.f. exciter is available which further reduces the noise level to less than 0·7% r.m.s. and increases the output power by 10%.

The laser and power pack together weigh 220 lb. Power input: 220 to 240 V, a.c., 50 Hz, 250 W.

Cooling: Natural convection.

SPECTRA PHYSICS

1255, Terra Bella Avenue, Mountain View, California 94040, USA. British Sales Offices: Queensway, Glenrothes, Fife, Scotland.

Model 125A helium-neon laser

This is the successor to the Model 125. It has a power of 50 mW at a wavelength of 6328 Å.

In its standard form the laser is d.c. excited and the noise level is better than 2% r.m.s. An r.f. excitation option is available which reduces the noise level to better than 0·3% r.m.s. and increases the power output by 5% to 10%.

Power input: 115/230 V, a.c., 50/60 Hz, 450 W.

Cooling: Natural convection

Model 141 argon ion laser

Power output of the Model 141 argon ion laser

Wavelength (Å)	Power (mW)
4765	30
4880	100
4965	30
5017	5
5145	100

The laser is r.f. excited and the noise level is less than 1% r.m.s. (120 Hz and 360 Hz).

Wavelength selection is accomplished by changing the output mirror, and a set of three mirrors is supplied with the laser to cover the 4880 Å and 5145 Å lines and all the lines at once.

Power requirements: 208/240 V, 3-phase at about 15 A per line, 50/60 Hz.

Cooling: Closed loop cooling system with filter from power supply to laser head. 2·5 gallons per minute, 25 psi hydrostatic pressure raw water required by the heat exchanger in the power supply.

Total weight: 415 lb.

SCIENTIFICA AND COOK ELECTRONICS LTD

40–48, High Street, Acton, London, W.3.

Scientifica and Cook have recently announced a 2 W argon ion laser which will have a sealed ceramic plasma tube in a thermally stabilised cavity. The laser tube is supplied by a highly stabilised solid state power supply.

SOLID STATE NUTRONICS LTD

5, Voltaire Road, London, S.W.4.

Argon ion laser Model NL 200

The guaranteed operational life is 1000 hours, or one year at 200 mW all lines output power.

The laser is d.c. excited and has a completely sealed tube. The various wavelengths are selected with a wavelength selector. The total weight of the power and the laser head is 248 lb.

Power input: 110 to 120 V, 200 to 250 V, 40 to 60 Hz, 13 A.

Cooling: Tap water at $1\frac{1}{4}$ gallons per minute and forced air.

Power output of the Model NL 200 argon ion laser

Wavelength (Å)	Power (mW)
4579	Weak
4764	Weak
4880	100
4965	Weak
5017	Weak
5145	60

CARSON LABORATORIES, INC.

375, Lake Avenue, Bristol, Connecticut 06010, USA.

Model 10 SP argon/krypton ion laser

Wavelength (Å)	Power (mW)
4880	70
5145	70
5682	70
6471	70

This is the only commercial mixed gas laser so far available. It is d.c. excited and after the initial warm-up period the output stability is better than ± 2%.

The output power is continually monitored and displayed on a meter on the power supply. The laser head weighs 100 lbs and the total weight of the system is 196 lb.

The various wavelengths are selected by a prism wavelength selector.

Power requirements: 195/210 V, 3 phase, 30 A, 50/60 Hz and 110 V, single phase, 10 A.

Cooling: Tap water at less than 15°C at 3–5 gallons per minute, 50–70 psi hydrostatic pressure to supply a heat exchanger.

R.C.A.

Gas Laser Marketing, Industrial Tube Products, Lancaster, Pennsylvania 17604, USA. British Sales Office: R.C.A. Ltd, Laser Sales, Tube Products, Sunbury-on-Thames, Middlesex.

The majority of laser manufacturers guarantee their products for either a year (regardless of hours use) or for 1000 hours use. R.C.A., however, have a different system under which they fully guarantee the laser against failure for 500 hours and on a pro-rata adjustment for 3000 hours.

Model LD 2100 Argon ion laser

The long term stability of this laser is better than ± 5% and the noise and ripple is less than 1·5% r.m.s. A prism wavelength selector is used to isolate the individual wavelengths. The total weight of the system is 260 lb.

Power requirements: 117 V, 16 A, 50/60 Hz.

Cooling: Air cooled by integral fans.

Wavelength (Å)	Power (mW)
4579	5
4658	1
4727	5
4765	16
4880	50
4965	10
5017	5
5145	37

Model 2127 Krypton ion laser

Wavelength (Å)	Power (mW)
4762	80
4825	10
5208	140
5682	60
6471	195

The laser is d.c. excited and it has a long term stability of better than ± 5% and ripple and noise of less than 1·5% r.m.s. A prism wavelength selector is used to select the desired wavelength. The total weight of the system is 1428 lb, of which the head contributes 395 lb.

Power requirements: 230 V, 50 A, 3 phase, 50/60 Hz.

Cooling: Water at less than 25°C at a rate of 2 gallons per minute and a pressure of 50 p.s.i.g.

Model LD 2101 Argon ion laser

D.c. excitation is employed and the long term stability is better than ± 5% while noise and ripple is less than 1·5% r.m.s. The various wavelengths are selected with a prism wavelength selector. The total weight of the system is about 1228 lb, the head comprising 290 lb of the total.

Power requirements: 230 V, 3 phase, 35 A, 50/60 Hz.

Cooling: Water at 4 gallons per minute and 50 p.s.i.g. minimum at an inlet temperature of less than 25°C.

Wavelength (Å)	Power (mW)
4579	50
4658	35
4727	35
4765	160
4880	650
4965	150
5017	80
5145	650

Appendix C

SOLVENT SPECTRA

All the spectra in this appendix were obtained on the Southampton University Chemistry Department's Cary 81 laser Raman spectrometer. The spectra were obtained using 180° illumination and are reproduced exactly as they were recorded. In each case the sample volume was about $70\,\mu l$ and the temperature was 24°C.

The spectra obtained at low sensitivity give a good indication of the Raman scattering efficiency of the solvents. In general, the best solvent for Raman spectroscopy will be the one which dissolves the largest quantity of solute and has the weakest Raman spectrum. The solvent with the weakest spectrum will be the one which required the highest sensitivity for its acquisition.

Water is the outstanding Raman solvent as its spectrum is so weak that it could not be recorded at all under the conditions used for the solvents in the spectra section.

The spectra obtained at higher sensitivity are included to indicate the kind of signal-to-noise ratios and the weaker solvent bands which could be encountered in dilute solution work.

In each case the polarisation state of the lines was determined by recording the spectra with the polariser and analyser parallel $\parallel$ and crossed $\perp$.

The exact positions of the lines measured in wavenumbers from the exciting line and their approximate intensities relative to the strongest line in the spectrum (taken as 100), are shown. The vertical arrows on the spectra indicate the direction in which the zero suppression was altered.

Values in parenthesis in the data given are for the high sensitivity spectra where these differ from those for the low sensitivity spectra.

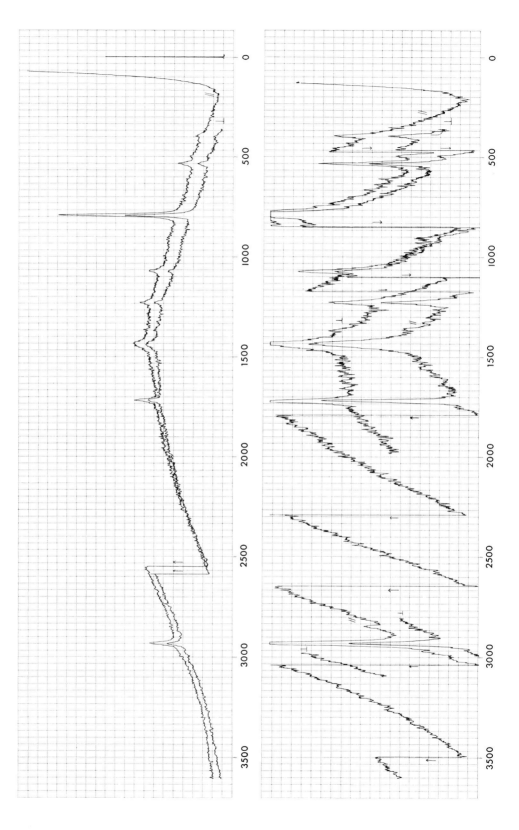

ACETONE

Source: BDH
Purity: Analar
Sample cell: Capillary
Reference dynode: 3
Raman dynode: 4
Spectral slit width: 4·6 cm^{-1}
Scan speed: 125 cm^{-1}/min
Slit length: 10 cm
Sensitivity: 400 (1500)
Single/double slits: single (double)
Pen response: 2 sec
Laser: He–Ne
Power: 40 mW

Δv (cm^{-1})	Relative Intensity
390	5
489	3
528	13
790	100
903	2
1068	10
1225	10
1366	3
1430	13
1707	18
2846	2
2922	21
2965	2

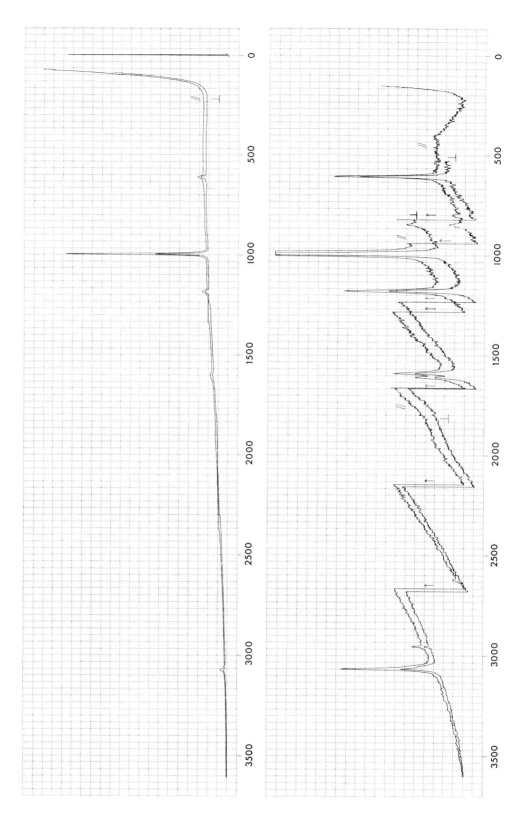

BENZENE

Source: Eastman Organic
Purity: Spectroscopic
Sample cell: Capillary
Reference dynode: 3
Raman dynode: 4
Spectral slit width: 4·6 cm^{-1}
Scan speed: 250 (125) cm^{-1}/min
Slit length: 10 cm
Sensitivity 34 (340)
Single/double slits: single (double)
Pen response: 0·5 (2) sec
Laser: He–Ne
Power: 40 mW

Δv (cm^{-1})	Relative Intensity
406	7
606	38
781	2
825	4
850	5
992	100
1180	31
1584	17
1605	17
2618	2
2949	5
3047	12
3063	29

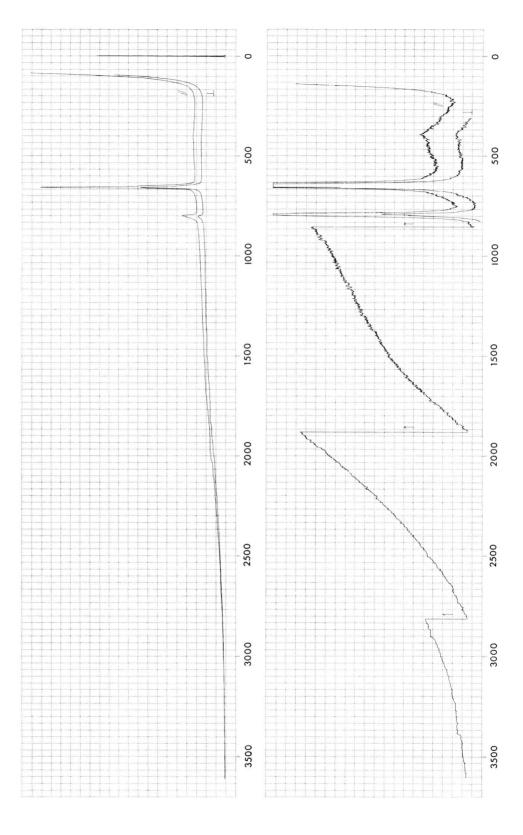

CARBON DISULPHIDE

Source: BDH
Purity: Reagent
Sample cell: Capillary
Reference dynode: 3
Raman dynode: 4
Spectral slit width: 4·6 cm^{-1}
Scan speed: 250 (125) cm^{-1}/min
Slit length: 10 cm
Sensitivity: 15 (150)
Single/double slits: single (double)
Pen response 0·5 (2) sec
Laser: He–Ne
Power: 40 mW

$\Delta\nu$ (cm^{-1})	Relative Intensity
397	1
648	36
657	100
796	7

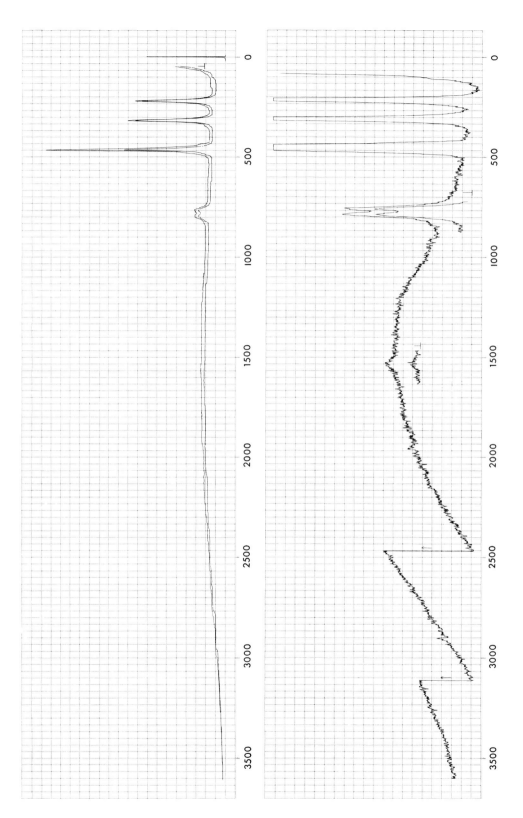

CARBON TETRACHLORIDE

Source: BDH
Purity: Analar
Sample cell: Capillary
Reference dynode: 3
Raman dynode: 4
Spectral slit width: 4·6 cm^{-1}
Scan speed: 125 cm^{-1}/min
Slit length: 10 cm
Sensitivity: 120 (600)
Single/double slits: single (double)
Pen response: 2 sec
Laser: He–Ne
Power: 40 mW

$\Delta\nu$ (cm^{-1})	Relative Intensity
218	46
314	54
459	100
762	8
791	8
1539	1

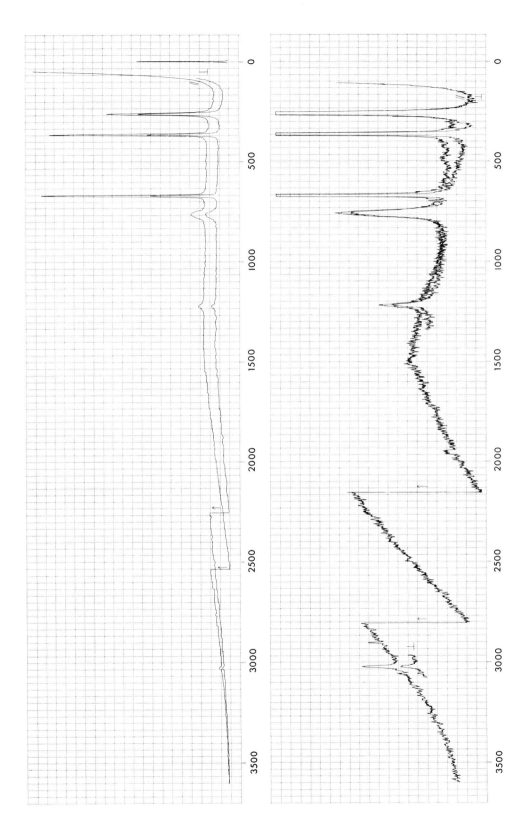

CHLOROFORM

Source: BDH
Purity: Analar
Sample cell: Capillary
Reference dynode: 3
Raman dynode: 4
Spectral slit width: 4·6 cm^{-1}
Scan speed: 125 cm^{-1}/min
Slit length: 10 cm
Sensitivity: 150
Single/double slits: single (double)
Pen response: 2 sec
Laser: He–Ne
Power: 40 mW

$\Delta\nu$ (cm^{-1})	Relative Intensity
261	64
366	96
668	100
758	13
1218	42
3024	2

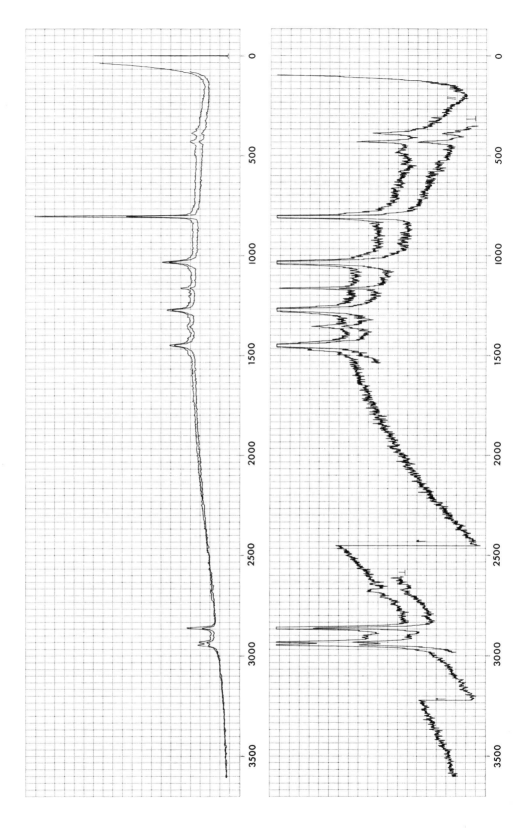

CYCLOHEXANE

Source: BDH
Purity: Spectroscopic grade
Sample cell: Capillary
Reference dynode: 3
Raman dynode: 4
Spectral slit width: 4·6 cm^{-1}
Scan speed: 125 cm^{-1}/min
Slit length: 10 cm
Sensitivity: 200 (1000)
Single/double slits: single
Pen response: 2 sec
Laser: He–Ne
Power: 40 mW

Δv (cm^{-1})	Relative Intensity
385	6
430	6
802	100
1030	25
1159	6
1266	12
1345	3
1446	12
1468	3
2630	1
2670	2
2697	1
2855	12
2905	1
2925	12
2941	12

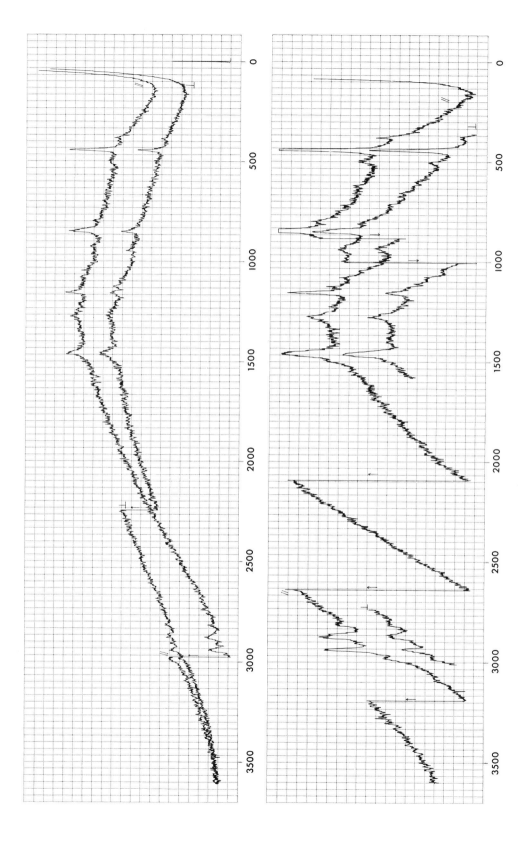

DIETHYL ETHER

Source: BDH
Purity: Analar
Sample cell: Capillary
Reference dynode: 3
Raman dynode: 4
Spectral slit width: 4·6 cm^{-1}
Scan speed: 125 (50) cm^{-1}/min
Slit length: 10 cm
Sensitivity: 500 (1000)
Single/double slits: single (double)
Pen response: 2 (5) sec
Laser: He–Ne
Power: 70 mW

$\Delta\nu$ (cm^{-1})	Relative Intensity
374	15
440	100
498	15
795	15
842	5
927	15
1031	5
1078	10
1122	10
1151	60
1273	25
1455	55
1485	10
2696	5
2806	15
2867	40
2898	20
2933	40
2979	10

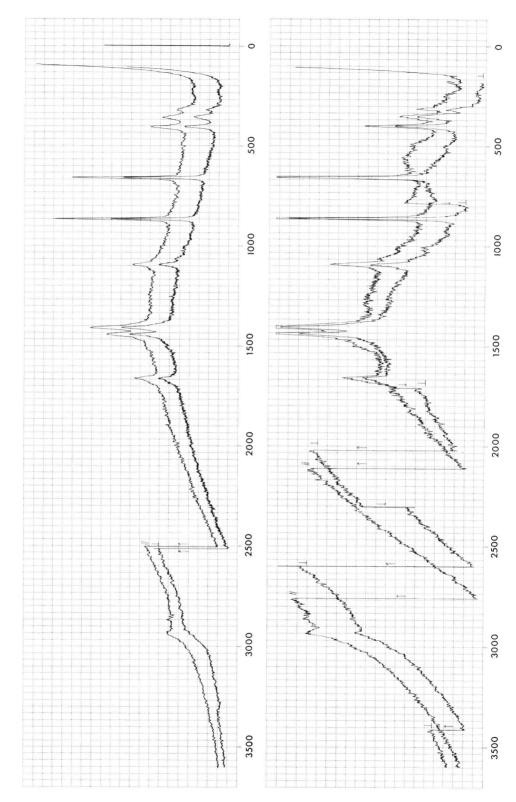

DIMETHYL FORMAMIDE

Source: BDH
Purity: Reagent
Sample cell: Capillary
Reference dynode: 3
Raman dynode: 4
Spectral slit width: 4·6 cm^{-1}
Scan speed: 250 (125) cm^{-1}/min
Slit length: 10 cm
Sensitivity: 200 (500)
Single/double slits: single (double)
Pen response 1 (2) sec
Laser: He–Ne
Power: 70 mW

$\Delta\nu$ (cm^{-1})	Relative Intensity
320	12
360	19
407	31
661	100
868	98
1065	2
1095	20
1407	52
1440	38
1662	18
2802	1
2860	3
2906	Spurious? 1
2932	6

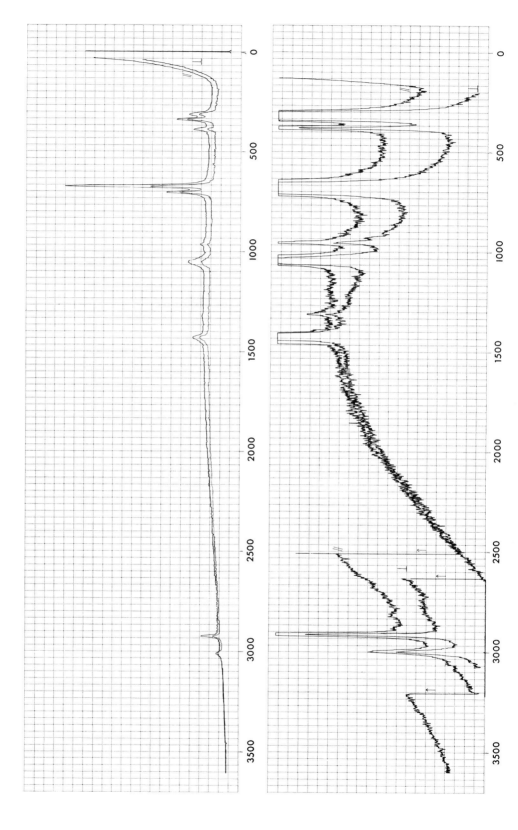

DIMETHYL SULPHOXIDE

Source: BDH
Purity: Reagent
Sample cell: Capillary
Reference dynode: 3
Raman dynode: 4
Spectral slit width: 4·6 cm^{-1}
Scan speed: 500 (125) cm^{-1}/min
Slit length: 10 cm
Sensitivity 85
Single double slits: single (double)
Pen response: 0·4 (2) sec
Laser: He–Ne
Power: 40 mW

$\Delta\nu$ (cm^{-1})	Relative Intensity
311	13
336	22
386	10
670	100
700	25
954	23
1045	12
1422	7
2885	1
2915	12
3002	6

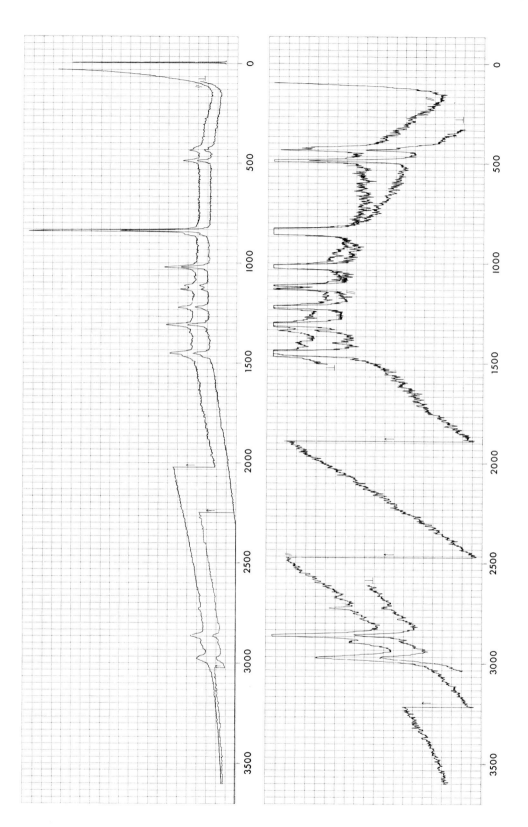

DIOXAN

Source: BDH
Purity: Analar
Sample cell: Capillary
Reference dynode: 3
Raman dynode: 4
Spectral slit width: 4·6 cm^{-1}
Scan speed: 125 cm^{-1}/min
Slit length: 10 cm
Sensitivity: 180 (1000)
Single/double slits: single (double)
Pen response: 2 sec
Laser: He-Ne
Power: 40 mW

$\Delta \nu$ (cm^{-1})	Relative Intensity
422	11
433	25
488	25
835	100
851	7
946	1
1015	19
1109	7
1125	7
1220	7
1308	15
1342	4
1403	4
1448	11
1460	4
2662	1
2720	2
2780	1
2858	7
2890	3
2968	7

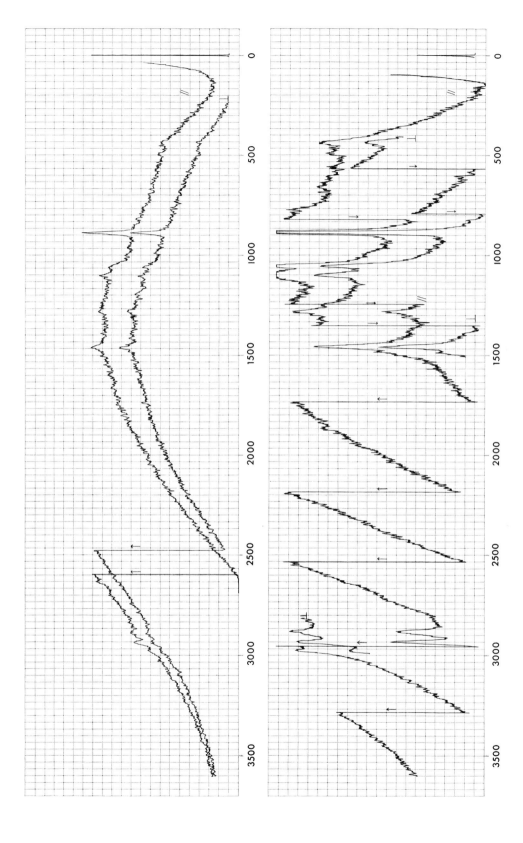

ETHANOL

Source: BDH
Purity: Analar
Sample cell: Capillary
Reference dynode: 3 (2)
Raman dynode: 4
Spectral slit width: 4·6 cm^{-1}
Scan speed: 125 cm^{-1}/min
Slit length: 10 cm
Sensitivity: 600 (100)
Single/double slits: single (double)
Pen response: 2 sec
Laser: He–Ne
Power: 40 (70) mW

$\Delta\nu$ (cm^{-1})	Relative Intensity
431	25
882	100
1047	20
1095	17
1274	8
1454	29
1480	17
2877	13
2928	21
2974	4

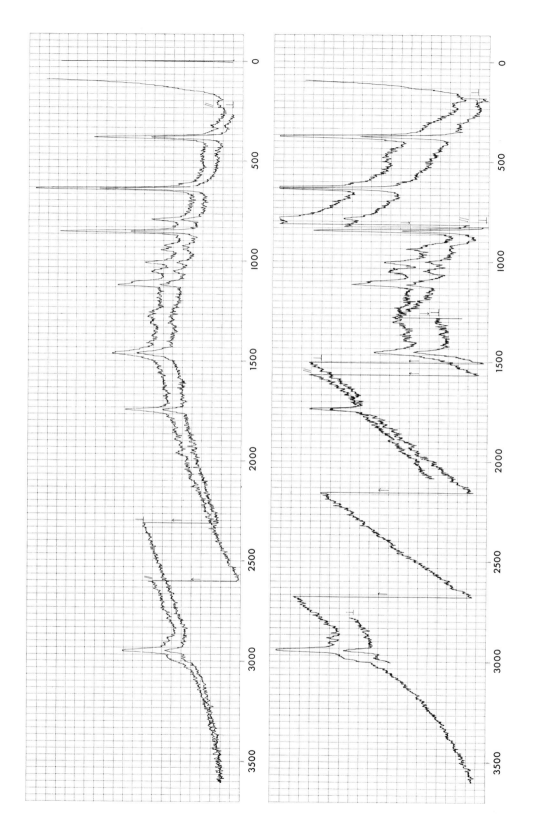

ETHYL ACETATE

Source: BDH
Purity: Analar
Sample cell: Capillary
Reference dynode: 3 (2)
Raman dynode: 4
Spectral slit width: 4·6 cm^{-1}
Scan speed: 125 cm^1/min
Slit length: 10 cm
Sensitivity: 500 (600)
Single/double slits: single (double)
Pen response: 2 sec
Laser: He–Ne
Power: 40 (70) mW

Δv (cm^{-1})	Relative Intensity
390	70
636	100
788	19
848	70
919	7
939	11
1003	15
1049	15
1103	18
1117	22
1457	18
1564	4
1739	18
2880	5
2906	7
2944	26
2970	3

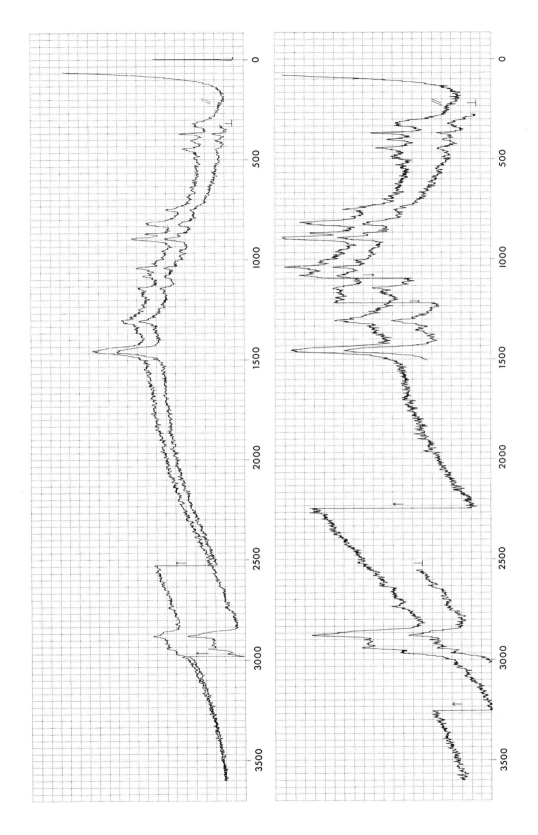

HEXANE

Source: BDH
Purity: Spectroscopic
Sample cell: Capillary
Reference dynode: 2
Raman dynode: 4
Spectral slit width: 4·6 cm^{-1}
Scan speed: 125 cm^{-1}/min
Slit length: 10 cm
Sensitivity: 500 (1000)
Single/double slits: single
Pen response: 2 sec
Laser: He–Ne
Power: 70 mW

$\Delta\nu$ (cm^{-1})	Relative Intensity
320	41
366	50
400	33
445	41
488	25
525	16
750	33
820	67
870	41
895	67
960	2
1012	16
1040	33
1068	8
1080	16
1142	25
1163	16
1308	41
1454	92
2625	8
2733	16
2875	100
2900	67
2920	67
2938	67
2960	33

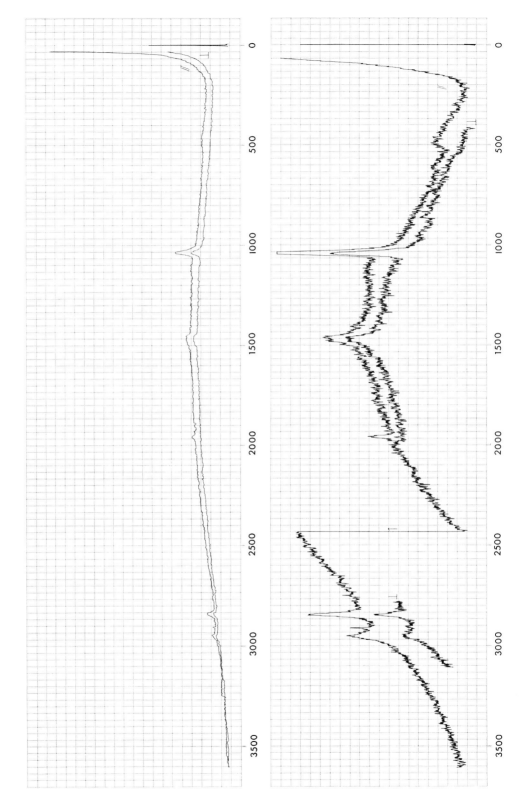

METHANOL

Source: BDH
Purity: Analar
Sample cell: Capillary
Reference dynode: 3
Raman dynode: 4
Spectral slit width: 4·6 cm^{-1}
Scan speed: 125 (50) cm^{-1}/min
Slit length: 10 cm
Sensitivity: 1000
Single/double slits: single
Pen response: 5 sec
Laser: He–Ne
Power: 70 mW

Δv (cm^{-1})	Relative Intensity
1034	100
1111	9
1164	4
1457	22
2838	45
2903	27
2947	27
3404	4

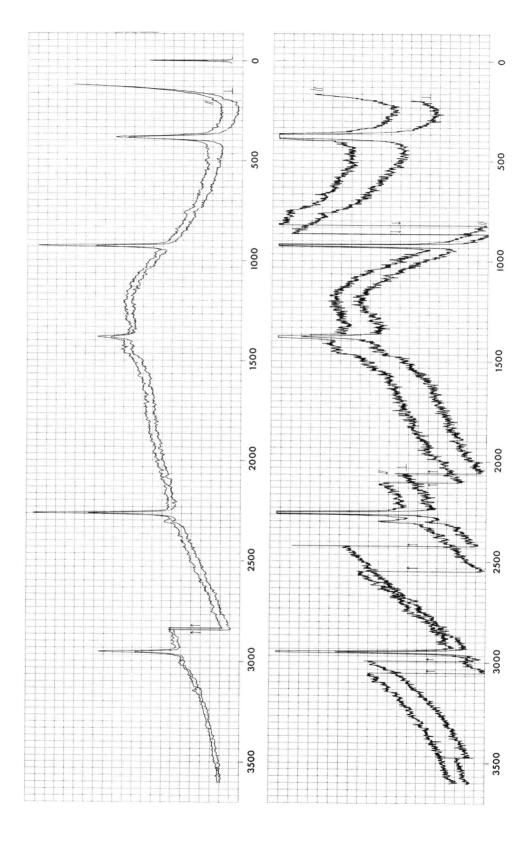

METHYL CYANIDE

Source: BDH
Purity: Reagent
Sample cell: Capillary
Reference dynode: 2
Raman dynode: 4
Spectral slit width: 4·6 cm^{-1}
Scan speed 250 (125) cm^{-1}/min
Slit length: 10 cm
Sensitivity: 280 (1000)
Single/double slits: single
Pen response: 1 (2) sec
Laser: He–Ne
Power: 80 mW

Δv (cm^{-1})	Relative Intensity
380	73
918	100
1375	20
1440	7
2252	93
2944	53
3000	7

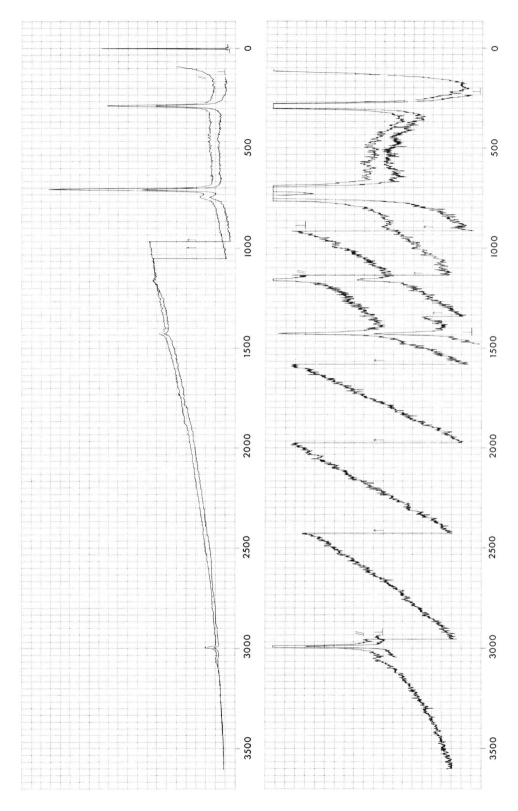

METHYLENE CHLORIDE

Source: BDH
Purity: Reagent
Sample cell: Capillary
Reference dynode: 3
Raman dynode: 4
Spectral slit width: 4·6 cm^{-1}
Scan speed: 250 (125) cm^{-1}/min
Slit length: 10 cm
Sensitivity: 100
Single/double slits: single (double)
Pen response: 1 sec
Laser: He–Ne
Power: 40 mW

$\Delta \nu$ (cm^{-1})	Relative Intensity
284	61
702	100
741	8
1156	2
1423	4
2985	8
3045	2

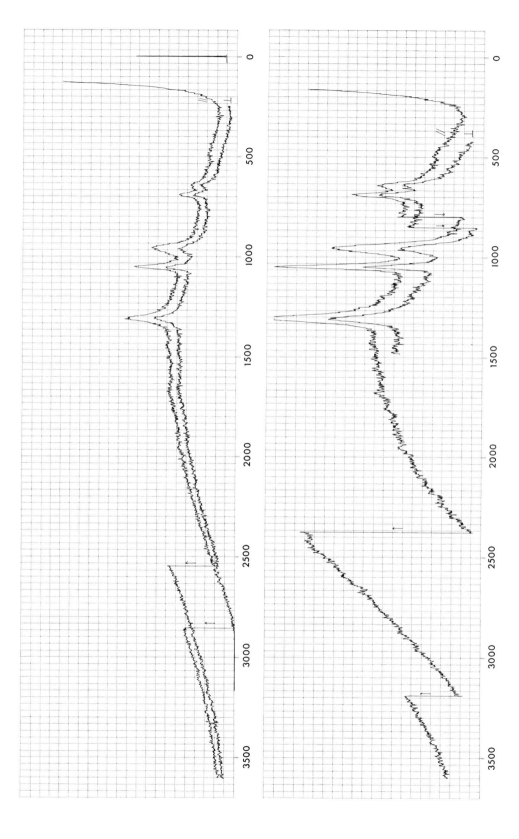

NITRIC ACID

Source: BDH
Purity: Analar–concentrated
Sample cell: Capillary
Reference dynode: 3
Raman dynode: 4
Spectral slit width: 4·6 cm^{-1}
Scan speed: 125 cm^{-1}/min
Slit length: 10 cm
Sensitivity: 500 (1000)
Single/double slits: single (double)
Pen response: 2 sec
Laser: He–Ne
Power: 40 mW

$\Delta \nu$ (cm^{-1})	Relative Intensity
470	1
650	20
690	40
956	80
1052	100
1313	100

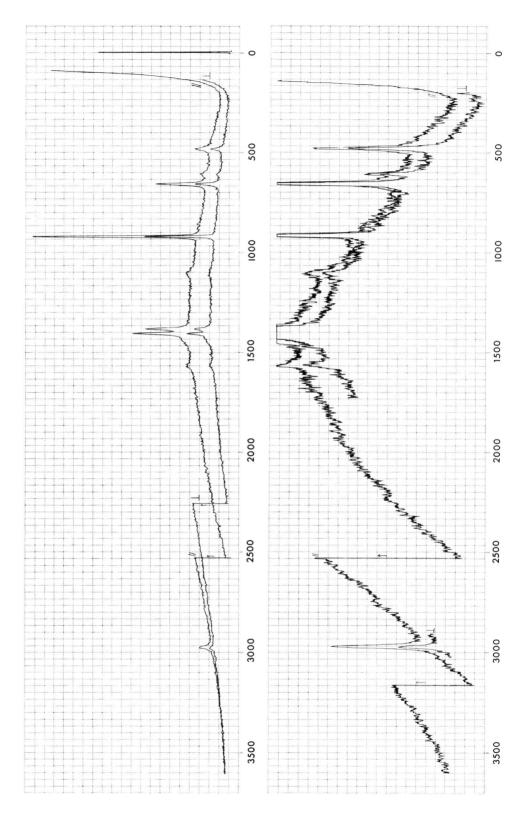

NITROMETHANE

Source: BDH
Purity: Spectroscopic Grade
Sample cell: Capillary
Reference dynode: 3
Raman dynode: 4
Spectral slit width: 4·6 cm^{-1}
Scan speed: 125 cm^{-1}/min
Slit length: 10 cm
Sensitivity: 240 (1200)
Single/double slits: Single (double)
Pen response: 2 sec
Laser: He–Ne
Power: 40 mW

$\Delta \nu$ (cm^{-1})	Relative Intensity
482	12
609	2
657	27
918	100
1105	4
1378	21
1401	29
1561	4
2766	1
2967	6
3038	1
3062	1

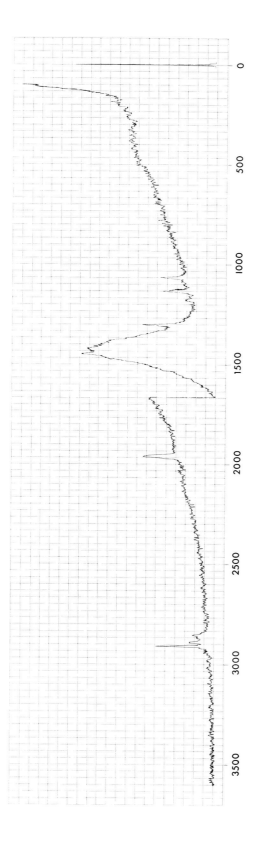

POLYTHENE

Source: ICI
Purity:
Sample cell: —
Reference dynode: 3
Raman dynode: 4
Spectral slit width: 4·6 cm^{-1}
Scan speed: 125 cm^{-1}/min
Slit length: 10 cm
Sensitivity: 1000
Single/double slits: single
Pen response: 2 sec
Laser: He–Ne
Power: 40 mW

$\Delta\nu$ (cm^{-1})		Relative Intensity
1065		46
1133		50
1174		18
1295		50
1415		14
1456		28
1954	spurious ?	65
2852		28
2882		36
2902	spurious ?	100
2941		7

Index